Wildlife Stewardship and Recreation on Private Lands

NUMBER ONE:

Texas A&M University Agriculture Series

Wildlife Stewardship and Recreation on Private Lands

By Delwin E. Benson, Ross "Skip" Shelton,
and Don W. Steinbach

Edited by Judy F. Winn

TEXAS A&M UNIVERSITY PRESS
College Station

The paper used in this book meets the minimum requirements
of the American National Standard for Permanence
of Paper for Printed Library Materials, z39.48-1984.
Binding materials have been chosen for durability.

Library of Congress Cataloging-in-Publication Data

Benson, Delwin E., 1949–
    Wildlife stewardship and recreation on private lands /by Delwin
E. Benson, Ross "Skip" Shelton, and Don W. Steinbach ; edited by
Judy F. Winn. — 1st ed.
        p.   cm. — (Texas A&M University agriculture series ; no. 1)
    Includes bibliographical references (p.   ) and index.
    ISBN 0-89096-872-1; ISBN 1-58544-445-6 (pbk.)
        1. Wildlife management.   2. Wildlife-related recreation—
Management. 3. Land use. 4. Right of property.   5. Wildlife
management—Economic aspects. I. Shelton, Ross. II. Steinbach,
Donny (Don W.) III. Winn, Judy F. IV. Title. V. Series: Texas
A&M University agriculture series ; no. 1.
SK355.B45   1999
333.78'15—dc21   98-47136
                                                              CIP

    ISBN 13: 978-1-58544-445-8 (pbk.)

Contents

Illustrations

Tables

Preface

The role of landowners in providing wildlife conservation, outdoor recreational opportunities, and ecosystem management on private lands in the United States is not well established in either theory or practice. The willingness of farmers and ranchers to include wildlife management and recreation in their economic enterprises and management plans is largely untried and uncertain. Some farmers and ranchers consider wildlife and the recreational potential it represents assets, both aesthetically and economically. Others have real or imagined problems with wildlife and recreationists and so may discourage animals, their habitat, and the people who seek them. This can lead to resentment on the part of landowners toward the agencies, organizations, and individuals who represent wildlife. At the same time, governments have struggled with the dilemma of managing public wildlife on private lands.

Solutions to this dilemma are both philosophical and pragmatic, but they must address the following questions:

- Are private lands important for producing wildlife and recreational opportunities?
- Is encouraging wildlife and recreation on private lands in the best interest of society?
- Will landowners benefit by managing for wildlife and recreation?
- Can the support of wildlife and recreation be a gratifying experience and a profitable component of agricultural life, or are wildlife and recreationists not compatible with production agriculture, livestock grazing, or forestry?
- How can landowners be encouraged to include wildlife and recreation in their management plans to help meet the growing demand for recreational opportunities based on wildlife resources?
- Will governments share some of their authority and responsibility with the private sector?
- Can the split estate of public wildlife ownership and private landownership become a partnership for managing wildlife on private lands?

In the United States most organized management for wildlife and the recreation associated with it is based upon public value systems, with policy formulated by local, state, and federal governments. However, two-thirds of the land in the United States is privately owned, and 80 percent of the lands that are economically feasible to improve for wildlife are under private ownership (Kimball 1963). Private lands are vital for conserving wildlife and for meeting the demand for recreational opportunities. Therefore, it is important for private lands and their owners to have a larger role in wildlife and recreation management. It is time for governments, organizations, landowners, recreationists, and communities to set common goals, share responsibilities, and work as partners in conserving and using our wildlife and other natural resources. If this book opens such a dialogue, it will have been successful.

The premises and perspectives upon which this book is based are the products of the authors' careers. We have had extensive tenure as professors and as Extension wildlife professionals. This educational perspective has allowed us to interact with our academic peers, personnel from natural resource management agencies, the landowning public, and consumptive and nonconsumptive users of wildlife. Dr. Delwin Benson has worked to bridge the gap between the needs of wildlife and the needs of those who produce and use wildlife on rural and urban private lands. He has international wildlife management experience in Canada, South Africa, South America, and several European countries. Dr. Ross "Skip" Shelton has a broad economic and wildlife management background, having worked and consulted with nongovernmental organizations, private landowners, and public agencies both nationally and internationally. Dr. Don Steinbach is a noted authority in the area of land-based recreational enterprises and has written about and done research in this area. Each of us has professional affiliations with state and federal land and wildlife agencies and has an appreciation of their roles and responsibilities in landscape and wildlife management. Our knowledge of the role of government in wildlife conservation has allowed us insights into the role of private land stewardship as well. We have observed that government does not have all of the solutions for wildlife conservation on private lands. The purpose of this book is to bring the private landowner and government into a mutually beneficial, positive relationship for the success of wildlife conservation.

Hunting is mentioned frequently in the book in the context of managing private lands for wildlife. Recreational hunting has a long history that yields practical examples of the way landowners and recreationists can interact positively. Those opposed to hunting should not conclude that the book has nothing to say to them. The ideas presented here have wide application. In the future we sincerely hope to see growing interest in nonconsumptive uses of wildlife to the benefit of landowners, the public, and wildlife habitat. We hope it will be clear to all that our intent is to foster constructive and optimistic dialogue among all those who care deeply about our wildlife resources.

Wildlife Stewardship and Recreation on Private Lands

CHAPTER I

Rights in Conflict

A thing is right when it tends to preserve the integrity, stability, and beauty of the biotic community. It is wrong when it tends otherwise.

—Aldo Leopold, *A Sand County Almanac*

There is no doubt that the integrity, stability, and beauty of the biotic community are at stake in North America and elsewhere in the world. In many overpopulated Third World countries the conversion of land to subsistence agriculture is occurring at an alarming rate. Land is cleared for cultivation or grazing with little thought to the effect on natural ecosystems and the wildlife they support. Individuals in such countries may use few resources, but their cumulative impact is monumental. In developed nations both virgin and agricultural open lands are disappearing because of rapid urbanization. The educated and economically elite inhabitants of developed countries are fewer in number than the Third World poor, but their individual and cumulative impacts on the environment are significantly greater.

In our materialistic society those who are able often seek high standards of living that put further strains on the land—second homes on the shores, in the mountains, or in rural communities; self-contained retirement communities; and small acreages where they can enjoy nature and open spaces with their families. Farmers and ranchers, whose families may have owned their lands for generations and who may have trouble supporting themselves with traditional agriculture, are often forced by high estate taxes or enticed by development dollars to sell their properties for development. As open spaces and "wild" landscapes are lost, so too are the animals that inhabit them and the access to those animals that so many desire. At the same time, governments are being pressured to provide more access to public lands, to protect wildlife on all lands—both public and private—and to preserve wilderness areas.

For some individuals the idea of wilderness with undisturbed wildlife is an abstract ideal because they rarely, if ever, seek access to it. Outdoor recreationists, however, may call for more government-operated parks and wilderness areas while demanding the kind of public access that puts such places at risk. Can governments satisfy public demands alone, and only on the lands within their jurisdictions? Can the private sector become a partner? Can we manage our rural areas so that wildlife and wildlife habitat are both protected and accessible? Can we find ways to make nature and wildlife competitive with development and complementary with agriculture? Given that private lands provide habitat for about 80 percent of our wildlife, it is clear that any attempt at managing and preserving wildlife must include sensible policies for making private lands and landowners part of the equation.

Who Owns Wildlife?

Who owns the wildlife in America? From the time of the Roman Empire, Western thought has placed wild animals in the same category as the air and the oceans in that they are considered the property of no one. Animals became the property of anyone who killed or captured them. The state had the authority to control the taking and use of that which belonged to no one in particular but was common to all (Council on Environmental Quality 1977). Private landowners had the right to protect their property from trespassing hunters and to maintain the exclusive right to hunt on their lands (Leopold 1933).

In feudal Europe the granting of licenses to hunt, formerly a government function, became intertwined with the right to own land and the right to control access to one's land. English laws and tradition were particularly important because of their later direct extension to America. The English Crown claimed the exclusive right to pursue game anywhere in the realm and could grant this right to others. After the signing of the Magna Carta, Parliament assumed the authority to grant hunting rights to the citizenry, but hunting was not normally open to the commoner. The aristocracy, through Parliament, developed a stringent code of laws to protect game and to make shooting the exclusive right of landowners. It was not until 1831 that the property ownership qualification was eliminated and hunting became possible for anyone who secured a license from the government.

It is easy to see the roots of American tradition in this history. Rebelling against the idea of government or a privileged aristocracy restricting individual rights,

4

Americans reestablished the ancient idea that wildlife is owned in common by all citizens. At the same time, however, the idea of private property rights became quite firm, allowing landowners to use their lands generally as they wished and to limit access.

In recent years the Supreme Court has reaffirmed the common ownership of wildlife and recognized the authority of states to regulate the taking of game animals. States are responsible primarily for resident wildlife, while the federal government deals with migratory, threatened, or endangered species. In carrying out this mission governments sometimes enact regulations that are not compatible with land management. These regulations, coupled with tax disincentives for setting aside agricultural land for wildlife, can discourage landowners from attempts at comprehensive wildlife stewardship. The assertion of governmental controls may have been necessary to deal with uncontrolled exploitation of wildlife in the nineteenth century, but by 1948 the Supreme Court was saying: "The whole ownership theory, in fact, is now regarded as but a fiction expressive in legal shorthand of the importance to its people that a state have power to preserve and regulate the exploitation of an important resource" (Tober 1981). So state ownership does not mean ownership in the sense of private property but rather custodial responsibility to preserve and regulate.

The central questions then become: How can governments carry out their mission of preserving the public's wildlife without encroaching on the property rights of private landowners? Given that most of our wildlife is found on private lands, what should be the role of the private landowner in managing and preserving wildlife and its habitat? Should landowners benefit from the presence of wildlife on their lands by selling access to recreationists? How can landowners manage wildlife and allow recreation on their lands without damaging their agricultural production or sacrificing their rural lifestyles? If governments discourage wildlife management and recreation on private lands, can public lands accommodate the growing number of recreationists without the destruction of wildlife habitat? And finally, can hunters, anglers, bird-watchers, campers, photographers, hikers, and other recreationists expect greater access to wildlife experiences without paying for it?

Wildlife Ownership and Property Rights

Perhaps it is time to explore new ideas that can bring all these vested interests together into a working partnership. Clearly, most governments do a good job of managing wildlife on public lands. Most private landowners also take seriously their responsibilities for land and wildlife stewardship. However, conflicts arise when various segments of society do not trust each other and when, as is often the case with environmental issues, viewpoints become polarized and extreme. Groups in conflict are less likely to cooperate when they perceive that issues are polarized and others are not listening. Sadly, this situation has existed in the United States in recent years.

One extreme holds that only the government can solve environmental problems and preserve our natural resources. Advocates of this view lobby for more laws regulating the activities of private landowners. Their argument is that private interests take away from public interests.

The other extreme mistrusts government and fears it will take away private rights through environmental laws and regulations. This may be a natural reaction for landowners who see their livelihoods threatened by regulations governing the ways they may use their lands. In the worst scenario, landowners might even be tempted to destroy wildlife and wildlife habitat to avoid public interest in their land and governmental attempts to control it.

We cannot afford either extreme, and neither will solve the complex demands on our environment. It is unthinkable that there should be no laws aimed at preserving wildlife and endangered species, and it is equally unthinkable that we should jeopardize the well-being of rural landowners and their opportunities for good wildlife stewardship.

We must face the future realistically. It is unlikely that governments will have the budgets to acquire vast new tracts of public land, and the public lands we now have are insufficient to preserve many wildlife species or to satisfy the demands of outdoor recreationists. Therefore, private land is very valuable to environmental interests and to recreationists. The problem is that wildlife, wildlife habitat, and outdoor recreationists are not always valuable to landowners. However, when landowners can benefit from the wildlife on their lands and the increased demand for access to wildlife and outdoor recreation, they will have an incentive to manage natural resources wisely. What is needed is a clear policy that responsibility for wildlife management should be shared by the public and private sectors. The goals of governmental wildlife managers, recreationists, landowners, educators, and business leaders should be to encourage such a policy and work together to put it into practice. There is a wealth of knowledge and experience in universities, private organizations, and government agencies that can help to make cooperation possible, but success will require a great deal of planning, trust, and goodwill, with sufficient social and economic incentives to compensate landowners for producing wildlife and granting recreational access to it.

Enfranchising Landowners as Partners

Enfranchisement is an official approach that the public and private sectors can use to begin organizing their objectives, roles, and responsibilities. Through this approach governments grant authority to private landowners to become wildlife and recreation managers on their lands, with the expectation that their stewardship will adhere to high standards. Landowners are helped and encouraged in that responsibility and are allowed to benefit socially and economically from their good management. In turn, landowners are expected to include wildlife and recreation more proactively in their plans and operations.

When individual landowners benefit, entire communities benefit as well. The protection of open space achieved by making wildlife and recreation economically and aesthetically valuable to communities is reason enough to invest in new approaches and partnerships.

To develop the idea of enfranchising landowners for wildlife stewardship, we will need to explore more fully the ownership of wildlife, the problems wildlife can cause for agricultural producers, the effects of historic and current wildlife policies, the importance of hunting in bridging the gap between public and private interests, and the ways nonconsumptive recreation may also become valuable. We will also take a look at trends in private wildlife management in the United States and elsewhere in the world—what works and what does not.

CHAPTER 2

Understanding
the Controversies

For our purposes, wildlife will be considered to be free-ranging vertebrate animals. These animals are not domesticated or confined in small enclosures.

In the United States wildlife are owned by all people. They are held in trust by the individual states, which share responsibility with the federal government for protecting migratory, threatened, and endangered species (Train 1978; Bean 1983). The state wildlife agencies deal with resident wildlife and enter into partnerships with federal agencies for managing species and lands within their control. The concept seems simple enough, but as Burger and Teer (1981) write, wildlife "is owned by both everybody and nobody [and—as a result] everybody's business is nobody's business. . . . What belongs to everybody belongs to nobody and is, therefore, fair game for anybody." Hardin (1968) calls this phenomenon "the tragedy of the commons." The tragedy becomes a management dilemma when wildlife are found on private land, which is protected by a different set of rights—property rights.

Private property rights in the United States include the rights to possess, enjoy, use, and dispose of economic goods, including one's land. Rights to land exist because governments recognize and enforce them. Trespass laws are one example of the recognition and enforcement of property rights. In exercising those private property rights landowners can choose how to use their land and its resources. Depending on their personal and business goals, they can choose to protect wildlife habitat or to alter it; they can choose to give the public access to the wildlife on their land or to deny access. Therefore, the public owns wildlife in theory but not in practice since most wildlife are under the control of private landowners. Unless private landowners are willing to provide habitat for wildlife and to grant access to the public, the public does not benefit from this resource. The conclusion is clear: Without productive collaboration with private land managers, governments cannot exercise effective responsibility for wildlife and the public cannot benefit from the wildlife it owns.

Experience Affects Attitudes

Almost everyone has some opinion about wildlife. While most of us are in favor of protecting and encouraging wildlife, attitudes can change quickly when animals cause problems. In urban or suburban settings woodpeckers make holes in buildings. Raccoons, squirrels, and bats move into attics. Pigeons, starlings, house sparrows, and geese nest and defecate in inconvenient places. Robins eat cherries from backyard trees, and squirrels harvest strawberries from gardens. Rats and mice destroy property and spread disease. Deer and elk eat ornamental plants in suburban yards, and hawks raid winter bird feeders in search of songbirds.

Animal damage to agriculture, however, often is not merely inconvenient but may also have significant economic consequences. Red-winged blackbirds and red-billed quelea destroy crops in the United States and Africa, respectively. Prairie dogs, elk, and African antelope eat vegetation intended for cattle. Elephants and hippopotamuses destroy subsistence crops. Coyotes, wolves, lions, leopards, and bears kill domestic livestock. Some larger predators kill people.

To survive economically, those who make their livings from the land must control animal damage that threatens crops and livestock. In the "unenlightened" past, that often meant trying to eliminate the entire population of a species. Gray wolves and grizzly bears almost met that fate in the United States. Certainly few people would want to see that happen to other species. The ultimate defense against problems with wildlife is, of course, to remove their habitat altogether. It may be in a landowner's best economic interest to drain a marsh to remove duck and blackbird pests, to stop growing alfalfa because it attracts deer and elk, or to clear sagebrush and other shrubs to encourage grass and to discourage deer. If wildlife damage cannot be controlled in a moderate way, habitat removal may appear to some to be the only alternative, although few people would condone that solution either. Yet we can understand the frustration of landowners when their reasonable attempts to control animal damage are thwarted by laws, costs, lack of effective prevention and control methods, and the attitudes of anticontrol groups.

The vast majority of landowners has no desire to remove all wildlife or all wildlife habitat from their lands. They, more than most of us in our highly urban society, value land and all the resources it contains. And most farmers and ranchers are adept at making their operations compatible with wildlife. However, if they are asked to become more intensive managers and to actively improve and increase wildlife populations and grant the public access to them, then society must be willing to help them in that role. Kellert (1981) notes that many landowners tend to view wildlife as yielding few if any returns and to be of minor importance compared to other resources. The potential of wildlife-based recreational enterprises has yet to be recognized by many.

One reason landowners may be reluctant to manage wildlife for recreation is their concern that hunters, or recreationists of any kind, may disrupt daily farming and ranching activities, pose liability risks, and cause damage or safety problems. They may believe that operating a recreational enterprise would require too much of their personal time and resources. Those are legitimate concerns which will be discussed in a later chapter.

Can agriculture and wildlife coexist? In most cases they can if wildlife contribute socially or economically to a supplementary and complementary enterprise, thus relieving the pressure on landowners to remove them. Historically, hunting has been the type of recreation from which landowners could most easily realize wildlife-based income. Sport hunters and their organizations have long been a force in wildlife conservation efforts. Hunters were the first to accept fee-based systems for access to private lands, and today the demand for such access is high. The demand for other types of wildlife recreation, including nonconsumptive activities, is growing and has potential for future economic benefits to landowners.

Past emphasis on whether the public or private sector had rights over wildlife access begged the question of how best to manage land, animals, and people. Drawing lines in the sand over rights may serve only to prevent meaningful debates and responsible actions. Landscapes and wildlife must be cared for. The public and private sectors must be responsible to future generations. Arguments about the importance of one right over another could jeopardize constructive opportunities for public and private sectors to work together toward a mutual conservation goal of long-term value. Landowners can be enfranchised to use and manage wildlife without giving their private property rights away.

History of U.S. Wildlife Policy

The uncontrolled destruction of wildlife for food and other commercial purposes and the removal of predators were recognized as problems and brought under control near the turn of the twentieth century. Laws were passed to prevent the overexploitation of wildlife, and the value of sustaining wildlife populations was recognized. President Theodore Roosevelt, along with sportsmen's organizations, was instrumental in pressing for legislation to protect large tracts of land. His wisdom and leadership set the stage for acquiring state and federal parks, wildlife refuges, forests, rangelands, and recreation areas. In 1909 Roosevelt forecast a combination of public and private contributions to wildlife conservation when he wrote about the need for both creating public reserves and intelligently managing private reserves.

A milestone in American wildlife conservation was the work of the American Game Policy Committee in 1929 and 1930. That group, led by chairman Aldo Leopold, developed the first wildlife policy to address the management of wildlife on private lands. Leopold suggested three things the government could do:

1. Buy private lands and become the owner;
2. Cede landowners title to the wildlife on their lands so that they could buy and sell them just as they would any other asset;
3. Compensate landowners directly or indirectly for producing a wildlife crop and for allowing others the privilege of harvesting it.

Leopold considered the first option feasible on cheap lands but prohibitive elsewhere. Government purchase of private land is still an option today, of course, but the inexpensive lands have already been acquired, and money for new purchases is difficult to obtain. Even when funds are available, state and local leaders are reluctant to remove private land from tax rolls, and conservative rural communities may be concerned about converting private land to public use. It is safe to assume that acquiring more public lands will continue to be a difficult and expensive option.

Today, as in Leopold's time, ceding the public's ownership of wildlife to private landowners seems inconsistent with North American values. In fact, wildlife management in North America has been firmly in the hands of government agencies. Yet, elsewhere in the world ceding rights in wildlife to landowners and rural communities has become an effective approach for preserving wildlife populations and emphasizing their value to society, as will be seen in chapter 4. Still, there seems to be a better way for this country.

Governments in North America generally selected Leopold's third option—providing incentives for landowners to produce wildlife, wildlife habitat, and recreation. Since then compensation to landowners has taken various forms in the United States (Benson 1988a, 1976). Owners have shared their lands by allowing access for recreation in exchange for help from government agencies with animal damage control and, perhaps, habitat enhancement advice and materials. Other compensation included paying for damage caused by wildlife and giving landowners permits to kill offending animals. States also provided hunting plans and the live removal of unwanted wildlife, which balanced the states' goals for animal numbers with the needs of landowners.

Educational programs in most states also helped landowners learn about recreation and wildlife management options on their properties. The Cooperative Extension System in some states took the lead in providing educational assistance to help landowners integrate wildlife management and recreation into normal ranch and farm operations. In most states wildlife and natural resource management agencies helped with habitat enhancement, but they generally avoided the promotion of fee hunting and other recreational enterprises.

Cooperation among agencies, hunters, and landowners began in the mid 1930s with pioneering programs such as Michigan's Williamston Plan, Pennsylvania's Cooperative Farm Game and Safety Zone Programs, and North Carolina's Farm Game Program. In the 1960s and 1970s such programs moved west with the Cooperative Hunting Program in Ohio, Operation Respect in Idaho and Colorado, Acres for Wildlife in Nebraska, a hunting unit program in Utah, and Feel Free to Hunt in Washington. These ideas were sponsored by state wildlife agencies in cooperation with hunters and landowners. Wildlife and habitat management and hunter access were the primary objectives of each state program. Landowners were asked to give hunters access to their lands and to set aside property for wildlife management in exchange for signs, plant materials, law enforcement, and assistance from government biologists. In time, leadership waned in many of these programs, and as a result they lost momentum or ceased to exist (Benson 1976). In retrospect, the programs were well intentioned but unsuccessful as long-term, sustainable solutions. They addressed symptoms of problems but not the problems themselves.

At times regulations have been passed in the United States that compensated landowners by giving them portions of the values (rights) resulting from wildlife and hunting on their lands. For example, in Wyoming hunters gave landowners coupons from hunting licenses, which landowners could redeem for money. Some states also gave landowners preference in obtaining licenses for hunting on their own lands. A few states allowed landowners to distribute at their discretion a portion of the total number of hunting licenses issued; landowners could charge a fee for these licenses or give them away. Today some states are allowing landowners to form private management co-ops and to participate in wildlife management and hunting decisions along with state wildlife agency personnel.

In 1930 Leopold's committee recommended that free hunting on private land was not in the best interests of wildlife management. The committee understood that hunters should pay for the opportunity to hunt so that landowners would have an incentive to sustain wildlife populations. More than sixty years later access for hunting is in a steady decline in the United States (Cordell et al. 1988a; Wright and Kaiser 1986; Brown et al. 1984; Guynn and Schmidt 1984). State incentives to ensure recreational access have met with limited success (Holecek and Westfall 1977; Barclay 1966; Johnson 1966). However, Wright and Kaiser (1986) state that the lack of adequate financial compensation may not be the most significant factor in landowners' decisions to deny access. A more serious problem is the improper behavior of hunters. One cannot expect landowners to accommodate recreationists who misbehave. Of all incentive programs, the one that makes the most sense allows landowners to be involved with decisions about managing wildlife and wildlife habitat on their lands and to have control over the actions of those who use their lands.

CHAPTER 3

The Commercialization of Wildlife

Although a controversial subject, the commercialization of wildlife may be the only way to promote better wildlife management on private lands. In spite of the early programs to compensate landowners for wildlife production and recreational access, the current system of wildlife management in North America has tended to focus on the acquisition of public lands, managed by public servants, to preserve the public's wildlife for the public good. But private landowners cannot be ignored; the public's wildlife affects them and is under their de facto control, and they too are part of the public sector. Therefore, the problems, rights, and influences of private landowners must be recognized, along with the tremendous potential that exists for ensuring the preservation of wildlife through better management on private lands. We have established that private lands and the wildlife they support have value to society. How then do we ensure that wildlife will have value to the persons whose lands they inhabit?

The answer may be found in simple economics. We must help those landowners who want to operate wildlife recreation enterprises do so successfully. Land operators who benefit from wildlife and recreationists will encourage them; if there is no benefit then landowners may not. Stark examples of this principle may be seen in parts of the western United States where elk herds compete with livestock on grazing lands. In Arizona and New Mexico 61 percent and 60 percent of ranchers, respectively, are asking the government to reduce elk herds ("Arizona Public land Policy" 1990; Latham 1989). Other ranchers benefit from the hunting access fees they charge, and they encourage elk herds. These landowners want wildlife on their properties, and so do the persons on waiting lists for hunting access.

For some people, the mere idea of commercializing wildlife conjures up horrible visions of ruthless slaughter when, unlike today, hunting was uncontrolled.

Our memories of the exploitative practices of the past are slow to fade, and that is good if it strengthens our resolve to preserve wildlife resources. But it is short-sighted to suppose that wildlife on private lands can be preserved unless they are an economic asset—or at least not an economic liability—for the landowner. Wildlife must compete with all other possible uses for land.

The commercialization debate is not about converting ownership of wildlife from the public to individuals. It is about finding reasons for landowners to want wildlife on their lands. Altruism and aesthetic value are not reasons enough to encourage wildlife if a living must be made from the land and if the presence of wildlife could hinder that economic necessity.

In reality, there are already many commercial enterprises funded by wildlife. State fish and wildlife agencies buy land, build buildings, hire staff, and manage wildlife with income from fees paid by hunters and anglers. Federal excise taxes from the sale of firearms, ammunition, and fishing gear bolster state wildlife agency business accounts. Recreation and conservation organizations charge for magazines, memberships, and banquets in order to run their businesses, and they solicit contributions for conservation. Sporting-goods stores sell firearms, fishing equipment, camping gear, and related merchandise. Nature stores sell books, clothing, birdseed, and binoculars to those who enjoy watching wildlife. Guides and outfitters take clients into the forests, fields, and marshes on ecotours. Commercial businesses profit from providing transportation, food, lodging, and other goods and services to wildlife recreationists. The natural resource departments of universities are funded by state and federal budget appropriations, grants from resource management agencies, and contributions from individual donors. Even the federal government's land and wildlife management agencies operate on the tax dollars of society. Surely landowners should not be criticized for operating fee-based wildlife recreational enterprises on their lands.

The debate about the modern commercialization of wildlife through hunting, live sales, or the production of meat and other products may be confused by misunderstandings about the role harvesting animals plays in successful wildlife conservation. The euphemism "harvest" evolved as a way to describe the planned removal of wildlife surpluses. Just as crops grown too close together or domestic animals raised in excess will not produce adequately, some wildlife species will not thrive if allowed to overpopulate their habitat. Animals in excess of the population supportable by a particular habitat constitute a biological surplus. Government wildlife management agencies regulate hunting according to the size of the wildlife surplus so that it will not add to the overall mortality of the population. Wildlife populations not thinned by hunting may be vulnerable to more deaths from factors such as starvation, disease, weather, accidents, and predators. Agricultural producers can relate well to the harvest concept because they must remove livestock and crops each year to make way for new animals, crops, and production.

Hunting may be a controversial activity opposed by some, but when controlled properly it is a natural tool of wildlife population management. The production and harvest of wildlife for recreation, meat, and other products is much higher in European countries than in the United States (Bubenik 1989). The human population is also considerably larger in Europe than in the United States, yet wildlife populations are thriving. This suggests that Europeans have a system that promotes conservation. In Germany the public has open access to many lands for recreational purposes. People may enter privately owned farmlands or forestlands provided they do not interfere with agricultural or forestry enterprises. However, there is not open access to hunting. Hunters must have permission from landowners, and most of them pay for access privileges. Animals killed become the property of the landowners (state or private), who may sell the meat to the hunters or to the general public. Most game management and game law enforcement are handled by private reserve operators and private gamekeepers. Losses from poaching are minimal. The system produces meat and employment and is self-sustaining without draining taxpayers. In effect, in Germany the commercial value of wildlife pays for the preservation of wildlife.

In the United States the acquisition of public lands that provide wildlife habitat is decreasing. The use of both public and private lands in ways that are incompatible with wildlife is increasing. Wetlands are drained at a rate of 45,000 acres each year in the United States (Chandler 1988). In the Great Plains 4.5 million acres of rangelands were converted to crops during the 1970s and 1980s (Laycock 1987). Between 1967 and 1975, 1.87 million acres of rural land were converted to urban areas, transportation corridors, or reservoirs each year. The use of pesticides, herbicides, and fertilizers has increased by millions of tons per year in recent decades. Timber removal increased by 27 percent between 1970 and 1986, while new plantings increased by only 0.8 percent (Hagenstein 1990). These factors have contributed to an enormous loss of wildlife habitat. If wildlife resources are to remain part of the fabric of our total landscape and environment, if this is really our "conservation ethic," then we must find ways to make the maintenance of habitat compatible and competitive with other land uses.

There is a strong and growing market for wildlife enterprises. Visits to national parks, forests, and recreation areas have increased by an average of 4 percent every year since 1977, and more people are traveling farther from home to visit these places (Cordell 1990). In 1996, 77 million Americans sixteen years old and older participated in some form of wildlife-related recreation (*1996 National Survey* 1997).

Cordell's survey of sixteen thousand private landowners in forty-eight states shows that on 48 percent of the private land hunting was restricted to persons acquainted with the owner (Cordell et al. 1988a). Only 23 percent of the land was open to the public. Yet 95 percent of landowners surveyed considered income from recreational access to their lands an important reason for owning land. In the East, Brown et al. (1984) report that approximately 50 percent of private land in upstate New York was posted against hunting. Guynn and Schmidt (1984) report that 79 percent of private land in Colorado was closed to hunting in 1977.

The 1996 National Survey of Fishing, Hunting and Wildlife Associated Recreation (Anonymous 1997) shows that 51 percent of the hunters in the United States used private lands only, and another 30 percent used public and private lands. Conversely, wildlife watching took place mostly on public lands (51 percent) or on public and private lands (34 percent). Just 10 percent of wildlife watchers used private lands exclusively. Benson (1988c) reports that in a random sample of 355 hunters in five states 51 percent were willing to pay for hunting access. According to the same study, 55 percent of National Wildlife Federation affiliate leaders, representing thirty-four states, expressed a willingness to pay for hunting. Berryman (1981) writes, "The plain average sportsman has been ahead of the professionals and the administrators, and has been willing for many years to accept the kind of system that will ensure the future of habitat and the opportunity to harvest game surpluses."

Clearly, merging the supply of wildlife on private lands with the demand for access to them would give greater commercial value to wildlife as an agricultural commodity and encourage good stewardship. The growing scarcity of wildlife-related recreational opportunities can be countered by cooperative programs between governments and landowners that enfranchise landowners to take a greater role in the production and sustained use of wildlife and habitat resources. In short, landowners would be given privileges for managing their lands well. Society would not give up its ownership of wildlife. Governments would not give up their custodial role of managing the public's wildlife. Hunting would still be regulated by state governments based upon wildlife populations and sustained production. Landowners would not be forced to sell access to recreationists, and recreationists would not be forced to pay for access; however, willing buyers and sellers would be brought together and assisted in becoming part of the conservation process.

The hunting and fishing community that helped form the conservation movement in North America has become discouraged about the current status of both conservation and access to hunting opportunities. Can wildlife conservation afford to lose historical allies and political champions of conservation? The 45 million hunters and anglers are an economic force that generated $59 billion in the United States during 1996 (*1996 National Survey* 1997). Resources on private lands can help support the existing demand and open new opportunities for the future if landowners have appropriate incentives. The same study revealed that 63 million wildlife watchers spent $29 billion. This growing group of outdoor enthusiasts could become the new allies of landowners when more of their recreational pursuits take place on private lands.

The commercialization and integrated management of wildlife on private land, with states and landowners as partners, would open a new resource base and constituency for wildlife conservation and should be the goal. No longer can private needs be overlooked. No longer should the resources on private lands be neglected. No longer should governments be afraid to deal with private landowners.

Today there are only a few strong state agency initiatives to help owners manage wildlife on private lands (see appendix D). Even at the federal level agencies such as the Cooperative Extension System, Natural Resources Conservation Service, and Farm Service Agency have relatively few staff members and programs that promote and encourage wildlife production, as compared to other uses of the land. If landowners were assisted to become better stewards of wildlife resources (and benefited economically from their efforts), they in turn would aid the conservation movement.

The issue is not whether wildlife should have commercial value, but rather whether their commercial value can help to preserve them and their habitat. The commercialization of wildlife for recreational opportunities, live sales, and meat production on private lands, under managed conditions, is compatible with modern wildlife management. This should be the goal and the new "frontier" for wildlife management professionals.

CHAPTER 4

Useful Models
of Landowner Enfranchisement

The African Experience

Persons in North America may not easily understand the different ways wildlife on private land might be managed. Therefore, it is useful to learn how people in other parts of the world manage their wildlife resources. The interactions between people and wildlife have some commonalities throughout the world.

When the welfare of people is enhanced by wildlife, then the people cannot afford to live without wildlife. If wildlife become a problem, people cannot afford to protect them. The real problems private landowners have to address are control of their own lands and the integration of wildlife and recreation into their agricultural and family plans. The following issues are at the root of the problem, and the solution:

1. How can wildlife and habitat management be pragmatically integrated with farm and ranch management practices without harming the economics of the agricultural enterprise or the lifestyle of the landowner and his or her family?

2. How can landowners grant access for recreation and still control their own lands and resources, as well as the actions of others? Landowners do not want persons on their lands who are not family, friends, or invited guests. If access is granted to the public through contract agreements or other means, they want it to be on their own terms.

3. How can landowners who want to manage for wildlife and establish recreational enterprises deal with the uncertainty caused by government regulations and the fear of financial liability?

We may find some answers by examining the way fee-based recreational programs and landowner-based management systems have evolved in Africa, where the situation is quite similar to that in the United States.

Rural lands and landowners in Africa and the United States have much in common. Some landowners are native peoples, while others are the descendants of immigrants from Europe and other parts of the world. In both places some lands are communal while others are privately owned. There is a strong focus on agriculture in rural areas, and many landowners in both parts of the world are trying to make their livings from the land. Given these circumstances, wildlife could be either conserved or abused.

The stories of how Africans are beginning to value and use wildlife on tribal and private lands can provide important lessons for North America. Farmers and ranchers in North America, with the "public's" wildlife on their lands, face similar philosophical and economic dilemmas as do Africans or, for that matter, owners of communal and private lands throughout the world. They all must ask how the needs of wildlife and the needs of people can both be met.

Ten factors are of greatest importance in understanding basic wildlife and recreation policy and management problems in Africa today:

1. Human population growth affects wildlife and wildlife habitat.
2. Cattle and agricultural crops affect wildlife and wildlife habitat.
3. Basic needs of humans are at stake.
4. Basic needs of wildlife are at stake.
5. Illegal use of wildlife is a problem.
6. Wildlife and recreation are not competing economically with other uses of natural resources (for example, farming and livestock production) at either the local or national level in most countries, with the exception of the Republic of South Africa.
7. Wildlife and recreation are usually inadequately represented in land-use planning and resource development schemes.
8. Wildlife and recreation may not be valuable to local peoples and may be detrimental to their economic livelihood.
9. There are too few professionally trained wildlife and recreation administrators, managers, and research biologists in most African countries. Most of the effort is toward managing wildlife and recreation in parks rather than on private and communal lands.
10. Governmental stability and administration are uncertain.

Do these factors sound familiar? The similarity with the current situation in North America is striking.

However, the political, economic, social, and natural resource systems in Africa are much more diverse than those in the United States. Some lands were set aside as parks and nature reserves during colonial times, with the remaining lands left in various forms of tribal, communal, state, and private tenure. Parks set aside for wildlife and recreation are often symbols of a country's developing environmental consciousness. But it is possible that the people of the country may not benefit from them. Even worse, when the people do not want parks, they may actually try to destroy park resources. A review of wildlife and habitat resources conducted for the International Institute for Environment and Development ([Teer] 1987) reported that although the number of African parks has grown sharply since the 1950s, some were little more than "paper parks" made up of lines drawn on maps but with no administration or protection. They were used improperly for grazing and poaching. About the same time, a sociological study in Tanzania reported that local antagonism toward parks was almost unanimous.

Philosophical arguments that wildlife are good, have intrinsic value, and have a right to exist are best accepted by affluent Western cultures. Persons who live at the subsistence level, and even those using wildlife commercially, have different thoughts about wildlife. Wildlife are to be eaten, kept from domestic crops and children, frightened from villages, and used for clothing, shelter, art, and religion. Wildlife that cause problems cannot be tolerated. Desirable wildlife are encouraged.

Ensuring that Wildlife Benefit Local People

The real concern in Africa is that local people will be displaced from their lands in the name of wildlife conservation. Noneconomic arguments that wildlife should be preserved for their intrinsic value are not totally practical or defensible from a human rights perspective. If people in undeveloped countries cannot satisfy their basic human needs or participate in decision making where wildlife resources are concerned, they will likely think it morally defective and elitist that some prefer wildlife preservation over satisfying those basic human needs.

In answer to this dilemma, Africans are evolving from the colonial-era idea of protecting natural resources with parks and nature areas (which disenfranchised local people who depend on those resources) to using economic enterprise strategies through which local people receive rewards from having wildlife and recreationists on their lands. North American management systems also began with government-centered approaches (i.e., parks, public lands, and public management agencies). While government agencies have managed public lands well, most wildlife and habitat resources are effectively out of their control. Pragmatically, governments must work with people, give them incentives for action, and empower them to implement conservation programs that governments cannot.

The obstacles to preserving wildlife in Africa are not easy to overcome. Eltringham (1984) writes that there is no hard evidence to justify wildlife preservation solely on economic grounds. Sport hunting brings revenue, but world markets may fluctuate. Meat sales have competitive markets and may not always be reliable income producers. He also warns that economic gain can lead and has led to overharvest of wildlife. This type of pessimism is healthy because it reflects a reality that must be incorporated into any wildlife management objective. Land and wildlife cannot be properly managed if all controls are lifted. The exploitative use of public resources is caused by the "tragedy of the commons" (Hardin 1968). If wildlife belong to no one, overharvest can result from a desire to harvest resources before one's competitor does. Harmful overharvesting of wildlife can be prevented if those who live on the land know that their basic living requirements will be met without exploiting wildlife.

Drury (1982) writes that African tribesmen must see that wildlife can provide value to their local communities. Tourism will play a role in areas where the infrastructure is adequate to accommodate tourists. Hunting and other sorts of wildlife consumption can contribute to local economies as well. It is far easier to advocate animal protection and conservation when local interests are being met and local peoples are involved in managing and sustaining wildlife populations. Decentralizing management plans gives local peoples incentives to care about conservation and to practice stewardship. The time has long since passed when the world's landscapes and wildlife can be preserved without having direct impact on large numbers of people, if not on entire countries. Tribal leaders, private landowners, wildlife conservation workers, and national leaders must develop sound plans, communicate them, and enforce them. It should not matter whether African countries have private or communal land tenure, or a combination, provided that governments assist in developing programs that include citizens in planning and then share the benefits of those programs with them. Decker (1987) recommends the prompt and visible distribution of benefits from wildlife and recreation management to the local people. Gilbert and Dodds (1992) state that wildlife must become economically valuable to both governments and local people if the resources are to survive. Nsanjama's (1993) conclusions about conservation of wildlife in Africa can also be the paradigm for North America:

> Conservation of wildlife in Africa must start with the premise that management of a resource includes deriving sustainable benefits for present and future generations. This means accepting that wildlife management must, in general, contribute to the productivity of other forms of land use, such as watershed management, agriculture, forestry, ranching, and fisheries. It must also provide food and other wildlife products, generate local employment opportunities and cash revenue in areas where there are no other sources, and enhance environmental stability. Wildlife management must protect aesthetic, scientific, cultural, and recreational values and must conserve the reservoir of genetic resources.

Over the past three decades southern African countries have emerged as leaders in establishing citizen-based conservation programs on both private and communal lands (Lewis and Carter 1993; Adams and McShane 1992), especially in Namibia (Joubert et al. 1983; Joubert 1974), Zimbabwe (Child 1995), and the Republic of South Africa (Benson 1991b).

Wildlife Ranching in South Africa

In the Republic of South Africa (South Africa), the term *wildlife ranching* refers to the production and use of animals on private lands. In South Africa most of the private lands are owned by Caucasians of European ancestry. Land ownership by rural blacks in southern Africa or elsewhere on the continent is more communal or tribal. Black cultures have been slow to establish Western style wildlife and recreation enterprises. Until recently these cultures saw wildlife more as a threat to human livelihood than as an economic or aesthetic resource. Several innovative approaches to tribal land management for wildlife and recreation will be discussed later.

The first laws that sought to preserve wildlife by restricting hunting were passed less than five years after the Dutch settled South Africa in 1652. Since then more than one hundred statutes and ordinances have been enacted to conserve wildlife. Conservation of wild animals is handled primarily by provincial governments, but several federal agencies and acts also are involved.

About 97 percent of lands in South Africa are outside nature reserves. Protection of habitat and wildlife outside nature reserves is one of the most difficult problems facing the conservation effort. This, undoubtedly, has been an impetus for South Africa's attempts to enfranchise landowners with authority to manage wildlife. Landowners have privileges over wildlife on private land. They are not required to have licenses to hunt ordinary game in the open season, and they may be granted licenses to hunt at other times of the year. Where landowners have established private reserves with fencing that adequately encloses animals, they have extensive hunting privileges. They may use certain firearms and methods that would otherwise be prohibited. Landowners may even be given the authority to control animal damage by removing individuals of protected species. Persons wishing access to private lands must have the permission of landowners. On some farms wildlife are produced for commercial purposes, while on others they are valued for aesthetics. Fee hunting is the commercial wildlife-based enterprise found most frequently on farms. Luxmoore (1985) provides evidence that conservation was improved in South Africa by offering profit incentives to landowners.

In the first formal countrywide study of the wildlife ranching industry in South Africa, Benson (1991b) discovered that the system was based on a positive relationship between public wildlife values and private land ownership and access control. He notes that all landowners, whether or not they sought income from wildlife, valued the presence of wildlife on their lands. Interestingly, those who sought income from wildlife (from meat sales, recreational hunting for meat, trophy hunting, live animal sales, or game viewing) valued wildlife more than those who did not and also were more involved in management practices that ensure the sustainability of wildlife populations.

In the study 81 percent of survey respondents reported that they fenced wildlife into large enclosures, allowing for private decision making about wildlife and recreation with the assistance of provincial wildlife agencies. Ranchers made 14 percent of their gross income from wildlife, spent 12 percent of their time with their wildlife enterprises, and had positive feelings about their efforts (94 percent). Wildlife ranchers believed that their neighbors approved of what they did (75 percent) and that their clients were happy (90 percent). The enterprises they ranked most important were wildlife meat sales, recreational hunting for meat, trophy hunting, live animal sales, and wildlife viewing.

The wildlife ranching system in South Africa is successful because of partnerships between the public and private sectors in which the authority and responsibility for wildlife and recreation management on private land are shared. Partnerships among landowners are also important. For example, the use of fences to enclose wildlife was important to the evolution of greater care and management by the private sector, but cooperatives among landowners that allow fences to be taken down will likely be the approach of the future.

Empowerment Elsewhere in Africa

Researchers from Colorado State University evaluated wildlife utilization models in South Africa and observed that successful programs had the following attributes: local people benefited; communities controlled resources and programs; local management was in place; there was independence from outside funding and interference; and there were low ratios of people to resources (Crook 1997). Some of the more successful programs show that local people and wildlife can have positive relationships.

Adams and McShane (1992) write, "Conservation has long operated on the comfortable belief that Africa is a paradise to be defended, even against the people who have lived there for thousands of years. The continuing reluctance to accept the link between vigorous indigenous culture and the survival of wildlife has led to conservation programs doomed to eventual failure because they depend on building barriers of one sort or another between people and wildlife. Such persistent blindness is tragic. . . . " In the Third World wildlife is not a luxury that only the educated and economically solvent take time to worry about.

There are programs that are bridging the barriers between people and wildlife. They do not all work equally well because of abuses at various administrative levels, but they are examples of legitimate attempts to empower local peoples (Adams and McShane 1992).

In 1960 Kenya started the most important early attempt to integrate the needs of local people with conservation efforts by reserving three thousand square miles along the eastern border of Tsavo National Park for the Wata, a tribe of renowned elephant hunters. The number of human residents in the area was limited to two hundred hunters and their families. They could hunt and sell products such as hides, meat, and trophies but not ivory, which was the property of the government. A quota of two hundred elephants could be killed each year. This program, called the Galana Game Management Scheme, failed because the government kept revenues from safari hunting rather than sharing them with the tribe.

The same idea was tried in Zambia in 1983 with the Lupande Development Project. Hunting and human settlement were allowed just outside the South Luangwa National Park. The project trained and employed local villagers, and some of the revenue from safari hunting was invested in community needs as identified by village leaders. This sharing of responsibility for wildlife management between local inhabitants and park employees led to a dramatic decrease in poaching rates. Village scouts, with their intimate knowledge of the area, were far more capable than park employees at patrolling and protecting the area. Scouts also provided data on wildlife densities, herd movements, trophy harvests by private companies, and crop damage from wildlife—data that had not been available before. While people in the project area were protective of wildlife, people elsewhere still believed that wildlife was of little benefit to them. Therefore, in 1987 the government expanded the program nationwide and renamed it the Administrative Management Design for Game Management Areas (ADMADE). ADMADE treats conservation as a business and villages as economic entities. As ADMADE expands, wildlife viewing is expected to bring in more revenue.

In Botswana the Chobe Enclave Project involves local people in hunting programs. The Department of Wildlife and National Parks sets quotas on the numbers of animals harvested by local people and safari hunters, and helps develop management plans. The department is changing from what once was a police force into an agency that gives technical assistance and training to local people so that they can better manage their lands. As communities become more responsible for wildlife conservation, the poaching pressure on animal populations in parks is decreased. Wildlife ranching on private lands is supported by the Botswana government as a way of giving wildlife economic value competitive with other possible uses of the land. The rationale is that the legal, commercial use of wildlife stabilizes the demand for wildlife products and can displace the illegal taking of game. Wildlife ranching requires that the landowner establish a secure form of property right or right of use, since a personal investment in the enterprise is required. Therefore, Botswana holds wildlife as a public resource, but the government can allocate custodial rights to landowners. Barnes and Kalikawe (1994) report that game ranching in Botswana is not profitable at the present time but that it has the potential to become so.

Bordering Botswana, in the Republic of South Africa, is the formerly independent state of Bophuthatswana, where wildlife use is contributing to community development. Davies, Grossman, and Rammutla (1994) report that parks and reserves are established only where they are the most beneficial use of the land and where there is popular support. The Parks Board manages wildlife in those areas. On tribal lands the policy is for a gradual and responsible shift of land ownership and management to local communities. Hunting is the most important income source to local communities. Local men are trained and employed as trackers, field experts, and camp workers. Tribal lands are still used for livestock production, but free-ranging wildlife are considered important as well.

Tanzania's Serengeti Regional Conservation Strategy has similar goals of helping communities adopt farming and hunting techniques that do not deplete natural resources and wildlife. Development assistance will bring clean water, sanitation, roads, schools, and health clinics to those in need while caring for natural resources over the long term.

Perhaps the best-known village wildlife program is CAMPFIRE—Communal Areas Management Program for Indigenous Resources—in Zimbabwe (Child 1995). It is based on the premises that people living with wildlife pay the price of conservation through injury and animal damages and so deserve to reap the rewards of conservation, and that people have the collective capacity to manage their natural resources. The plan envisions a system of natural resource cooperatives functioning in much the same way as private owners of commercial ranches; profits from natural resources enterprises may be used to benefit the community or dispersed to individuals. Although the program has been successful financially and ecologically in the Nyaminyami district, the concept has not yet gained full participation throughout Zimbabwe. Few people outside the district council level participate in discussions about policy and how money will be spent. Many local villagers still do not believe that wildlife belong to them, and most still see wildlife as a liability rather than an asset because of earlier colonial policies and the continuing competition between agriculture and wildlife. The problems will be solved when significant revenues reach the level of individual householders and farmers and they can begin to see the benefits of wildlife stewardship. About the CAMPFIRE program Taylor (1994) writes, "Wildlife management is as much an institutional problem as it is a technical one and its successful implementation lies in the hands of local people who will make the ultimate decision as to how they finally use their land."

Zimbabwe also has strong private landowner initiatives to accompany the management of parks, reserves, and tribal lands. Du Toit (1994) reports that with declining government investments undermining wildlife conservation on public land, the wildlife industry on private lands is stimulated by international ecotourism and by decreasing incentives for cattle production in low-rainfall zones. Some landowners manage independently, but many are forming conservancies that combine lands into extensive wildlife complexes. Adjacent tribal lands benefit from the increased wildlife on these conservancies. This approach joins landowners in planning and administrative functions and is a more ecosystem-based approach to management. Endangered black rhinos have been moved to Zimbabwean game ranches because these conservancies now offer better protection and management than government lands. Wild dogs and cheetahs also are managed cooperatively through the conservancy concept. There is a lesson here for the management of endangered species in America. Rational conservation will be served when wolves become an asset to landowners in the Yellowstone ecosystem and when black-footed ferrets are seen as ecotourism assets rather than economic liabilities for farmers and ranchers.

The common denominator of these successful African wildlife management programs on private and communal lands is, of course, the economic benefit of managed hunting. While wildlife viewing and other nonconsumptive types of recreation will surely grow in importance, it is the income from hunting that has financed these conservation efforts. Ironically, the very enterprise that is helping to save wildlife is criticized by some in the international conservation community. The influence of the animal rights movement in Europe and the United States could jeopardize programs that are making village wildlife management and conservation possible in Africa. In a report to the World Wildlife Fund, Freese (1994) states, "The question is not whether or not to use wild species, but rather how do we move from a system and level of use that is clearly not sustainable . . . toward a system that is better." Adams and McShane (1992) write, "Conservation needs a new cast of characters, drawn from both government institutions and the rural communities." Now is the time to strike a balance between government and private management to ensure the conservation of wildlife and natural resources.

CHAPTER 5

Enfranchisement Examples in the United States

There are also models of landowner enfranchisement in the United States. Let us look first at a few current programs in the western United States, where the abundance of public lands encourages governmental agencies to take charge of wildlife and recreation management. It is in the West that we see most clearly the controversies surrounding management of the public's wildlife on private lands and ways these controversies are being resolved so that better conservation can occur than either the public or private sector could achieve independently. Then we will examine programs in Texas that are somewhat representative of approaches to landowner enfranchisement in the southern United States.

Incentive Programs in the West

California has a history of fee hunting dating back to the 1800s (Arha 1996; Fitzhugh 1989; Mansfield et al. 1989). The current Private Lands Wildlife Habitat Enhancement and Management Area Program, approved by the legislature in 1979, provides incentives to landowners to better manage their lands for wildlife habitat. Landowners and wildlife agency officials reach broad agreement on a harvest program, then develop management plans based on the land area, land use practices, vegetative types, wildlife resources, historic and suggested harvest levels, management objectives, and proposed habitat improvement practices. Plans are reviewed and approved by agency personnel and are subject to periodic review. The landowner operates under the plan and is not restricted by general wildlife regulations such as season structure and timing or limitations on the sex of animals harvested. Management plans and licensing are for a five-year period and can be extended. The program peaked in 1989 and currently includes forty-six ranches comprising 545,139 acres. It has not grown recently because of changes in regulations and administrative procedures which cause uncertainty for land-owners. The wildlife agency defends the program on the basis that a substantial part of valuable wildlife habitat is under predictable management.

The Ranching for Wildlife Program in Colorado (Davis and Benson 1994) began in 1986 as a three-year experiment based on California's program. The goals of the Colorado Division of Wildlife (CDOW) included the usual recreational access, as well as habitat and wildlife population enhancement and improved relationships among agencies, sportsmen, and landowners. By 1989 other goals had been added: minimizing the cost to the CDOW and gaining public acceptance. A cap of twenty-five ranches was placed on the program in 1992, and in 1994 there were twenty-two ranches participating. To participate, a ranch must have twelve thousand or more acres in one contiguous unit, and title to the property must be held by an individual, corporation, or association. For each hunted species, at least 50 percent of the population must be on the property during the nonhunting season. Applications, management plans, and approval procedures are similar to those in California. A percentage of the total number of hunting licenses issued goes to landowners in the program. They may resell them or give them away. The remaining licenses go to the public through a drawing, and the hunters selected may hunt on private ranches in the program at no cost.

An evaluation of the program revealed that landowners, hunters, and agency personnel consider it successful because it creates incentives for private land-owners to produce wildlife and because it provides high-quality hunting for public and private hunters (Davis and Benson 1994). Hunting is generally better on ranches in the program than on public or private lands not in the program. The evaluation also showed that land operators in the program are committed to wildlife management. The technical assistance landowners receive is a strong point of the program. Unfortunately, because the number of ranches allowed to participate is so small, the program is not having as much positive impact as it could if allowed to grow. Public users (91 percent) were either "strongly," "somewhat," or "slightly" satisfied with their 1994 hunting experiences (Lloyd et al. 1996).

The Ranching for Wildlife program reestablished the almost vanished practice of free public hunting on the franchised ranches in the program. Based on the appraisal of landowners' management practices, habitat seems to be improving on those lands in the program. All program-related activities are funded by the landowners or their clients who pay the hunting fees. Landowners also invest in habitat improvements because they have the aesthetic and economic incentives to do so.

The CDOW is achieving its wildlife population objectives through appropriate harvests on franchises in the program, and if more land were included, native wildlife populations might be restored to a level of abundance and health not seen in Colorado for generations. One way to expand the program and bring more habitat under proper stewardship would be to encourage groups of ranchers to join the program as cooperating partners forming one entity, with the stipulation that they would develop a single management plan and appoint a single spokesperson to deal with the CDOW.

Some problems with the program must be overcome, however. Many wildlife conservationists believe that the public trust can only be served by public management. They are so convinced that private land management interests are at odds with wildlife management objectives that it is difficult for policy makers and bureaucrats to accurately appraise the potential of a program that transforms wildlife from a by-product to a primary product, or at least to an integrated product of land management. Because of this, the CDOW has shown some tendency to want to micromanage each ranch in the program, and such excessive control can become burdensome to both parties. As a matter of principle, it is more important to identify the incentives that will lead to good performance than to perfect command and control strategies. If the CDOW can find the correct set of incentives and requirements to guide the franchises, it can devote its scarce resources to monitoring and information management, technical assistance, cooperatively setting annual seasons and harvest quotas, and the periodic review of agreements, all of which are manageable burdens. These functions could be more effectively performed if wildlife management plans were more specific regarding the timing and location of habitat improvement practices, and if management practices were more directly linked to achieving herd management goals.

A second problem is the length of the franchise agreement, which is currently five years. It is reasonable for new ranches joining the program to be reviewed after three or four years, but after a satisfactory trial period an agreement could be written for a minimum of ten years. Considerable time is required for wildlife and land management practices to pay off, and landowners need this long-term security to encourage their financial commitment.

New Mexico began its Private Land Allocation System in the 1930s (Gonzales 1989). Neither detailed application procedures nor specific management plans are required for landowners to participate. The allocation of public and private licenses to hunt on private lands is based upon the amount of land in the program and the use of animals on those lands. Landowners may purchase a portion of the permits issued for their own use or for resale; the public may purchase the remaining permits and can hunt at no cost on the private ranches. The program reduces trespassing problems and benefits local economies through additional jobs and services. The New Mexico Game and Fish Department considers the program successful because game populations have been restored and wildlife management enhanced. Perhaps this program also suggests that governments do not have to micromanage landowners by insisting on detailed management plans and closely monitoring what is being done on private lands.

Utah's Posted Hunting Units Program (Arha 1996) began in 1939 as a pheasant hunting cooperative and was expanded in 1990, on a trial basis, to include big game. Its objectives are similar to those in other states: to secure and manage productive wildlife habitat and animal populations on private land; to improve public access to private land; to compensate landowners for their production of wildlife, habitat, and recreation through market-based economic incentives and enabling government regulations; and to improve relationships among landowners, sportsmen, and wildlife agencies. Except in special circumstances, landowners must possess at least ten thousand contiguous acres. Boundaries must be well marked, and hunting must be allowed on 75 percent of the hunting unit. As of 1994 forty-five big-game units were participating, comprising more than one million acres of land. The wildlife agency issues permits to hunters either through public drawings or on request of unit landowners.

These programs in western states are demonstrating that governments can enable landowners to deal with their own problems and encourage them to do what is correct in terms of wildlife stewardship. The needs of the public and private sectors are being met.

The Texas Lease System

States with little public land, primarily in the South, developed privately operated fee hunting and other fee-based recreation out of necessity. One of the best developed of these systems is in Texas, where hunters purchase the right of access to a parcel of land for recreational hunting over a specified period of time. The state is responsible for establishing hunting laws and regulations. Landowners may be individuals, partnerships, estates, or corporations. Hunters may be individuals or a group of individuals who act as a single legal entity. The three vested interests involved—private landowners, hunters, and the state—all influence the process, and the system cannot work successfully without the support of each of these interests.

The fee system did not develop uniformly in Texas; rather, the phenomenon is evolutionary. Yet, while the contractual agreements that result from negotiation between landowner and hunter are diverse, they contain common elements that, over time, have proved essential to the success of such arrangements. To understand these elements and the system itself, we must know a little of the history of fee-based hunting in Texas.

Early in this century game populations had dropped to all-time lows as a result of uncontrolled hunting and habitat destruction. At the time the state was struggling to establish an effective wildlife conservation agency. A few wildlife protection laws were passed (e.g., making the sale of wildlife unlawful), but for many years the only officers charged with enforcing the laws were county sheriffs who could do little. The first attempt to protect wildlife at the state level was the creation of the Texas Game, Fish and Oyster Commission (TGFOC) in 1907. The commission employed only six game wardens to patrol the whole state, 26,500 square miles (Gardner 1998). Alfred Gardner (1998), one of the original six game wardens, notes, "Without the cooperation of the ranchmen a game warden can do but little."

The development of roads and automobiles in the 1910s and 1920s made it possible for large numbers of hunters to reach the remotest wildlife ranges. Unrestrained and illegal hunting prevailed. W. J. Tucker (1943), executive secretary of the TGFOC in the 1920s, concludes, "Nothing else had so rapidly contributed to the decline of game in Texas since the heyday of market hunting." While game laws forbade the sale of game, defined seasons, and set sex and bag limits, they made no provision for controlling the distribution of hunters. In addition, it was practically impossible to enforce even these laws. According to the reports of the TGFOC of 1919 and 1924, there was little public sentiment for controlling hunting, and the state legislature declined to appropriate monies for hiring wardens. In addition, trespass laws were weak. To post land against hunting a landowner had to take civil action through the county sheriff's office. Under the laws of the times, any bona fide traveler, while traveling along a public road in a fenced pasture, had the right to kill game within four hundred yards of the road (Tucker 1943). Neither landowners nor the state had functional control of access to land for hunting. However, a few state district judges in the southwest part of the state gave sympathetic support to landowners by issuing blanket injunctions against all unauthorized hunting on certain ranches. In addition, some ranchers hired men to patrol their lands to keep out hunters (Gardner 1939; Tucker 1943).

During this period of excessive hunting, some hunters, in an effort to preserve hunting for themselves, surprised landowners by offering to pay them for the privilege of hunting on their properties, provided the landowners made an effort to curb illegal hunting. According to Tucker, "When leasing began, Texas had a strong law against the sale of game, and the sale of an exclusive privilege to hunt closely bordered on a violation, if it did not actually breach this statute. Therefore, most trades were surreptitiously made" (Tucker 1943). But the concept quickly grew in popularity. The TGFOC reported in 1922 that more and more landowners were limiting the hunting on their lands to themselves and those to whom they gave or sold hunting rights. The commission stated, "Each year sees more farms posted and fewer free shooting grounds, so some thrifty landowners are making the game render a direct return by selling or leasing the hunting rights. Many ranch owners in the Southwest (Texas) whose lands are stocked with deer are making more money from selling or leasing hunting rights on their ranches than they receive from leasing for pasturage."

In 1925 the state passed the Game Preserve Law, an official effort to preserve game by stopping hunting on some lands. Individual landowners could enroll their lands with the Game, Fish and Oyster Commission and have them designated "game preserves." Hunting was prohibited on such preserves, and game wardens could arrest hunters on officially enrolled lands. A second game law, also passed in 1925, was of greater significance to the development of hunting leases. The Shooting Preserve Law of Texas required all those who lease or sell hunting privileges on their lands to first obtain a shooting preserve license. The licensee was required to keep records on hunters and the animals they killed, and to enforce state laws. The confusing questions of game ownership and land ownership seem to be reflected in the terminology of the law, because a license, by definition, grants permission to do something that otherwise would be illegal (Fambrough 1987). TGFOC commissioner Tucker (1943) writes, "The motive for the . . . law was not . . . a recognition of the merit of the profit incentive to landowners as a means of protecting and increasing game, but merely for revenue purposes and to satisfy official curiosity as to the extent of hunting-ground leasing in Texas." A trespass law passed in 1929 prohibited a person from entering enclosed land for the purpose of hunting without first obtaining the consent of the landowner (Hill 1976), which strengthened landowner control of hunting.

Thus, hunting leases began during a time when poaching, dwindling game populations, and trespassing were concerns of landowners. Likewise, hunters were concerned about the shrinking game supply and any restrictions on their opportunities to hunt. The government faced the problems of preserving wildlife, improving game law enforcement, and providing equal opportunity to hunt. Thus, the three crucial groups—landowners, hunters, and state agencies—found their interests coming together. No single group pushed the others into participation, but each was drawn in by the circumstances of the times. The system has worked successfully and has spread to all areas of the state (Teer and Forrest 1968; Thomas et al. 1984). Hunting-lease income to Texas landowners rose from an estimated $13 million in 1965 (Klussmann 1966) to a range of $100 to $300 million in 1987 (Steinbach et al. 1987). At the same time, game populations have increased as more landowners find reason to preserve wildlife and wildlife habitat on their lands and to market hunting access to those lands.

The uniqueness of business-based systems in the southern United States and the Republic of South Africa is that they have a reasonably long history of sustained success compared with the compensation-based programs in other parts of the United States. The programs work because they deal with the basic needs of the land, landowners, wildlife, and recreationists. Landowners are solving their own problems and benefiting from their initiatives. Recreationists value and are willing to pay for places to enjoy hunting and other outdoor recreation. Conservation is enhanced because rural lands are managed for wildlife rather than against wildlife. Wildlife have not only aesthetic value but also pragmatic value, because they have become part of the economic productivity of the region.

CHAPTER 6

From Theory to Practice

Constraints to Landowner Enfranchisement

The transition from theory, history, and philosophy to concrete actions that can lead to the enfranchisement of private landowners for wildlife stewardship begins with a look at factors that currently constrain such enfranchisement.

Uncertainty

Landowners dread uncertainty, and uncertainty discourages long-term planning and investment (Shelton 1982). One source of uncertainty may be a landowner's unfamiliarity with the economics of wildlife management, which takes time to learn. Most management practices that lead to habitat and wildlife improvement have deferred returns. The same can be said of any agricultural enterprise. In order to approach wildlife management realistically, landowners should expect planning and implementation to require several years, with returns beginning to be realized only toward the end of that cycle.

Landowners interested in deriving income from wildlife need adequate financial data for habitat improvement in order to properly evaluate the potential for wildlife-based recreational operations. Potential economic benefit can certainly be an incentive for landowners to manage for, rather than against, wildlife. However, landowners cannot expect to derive significant returns from hunting leases or any other recreational enterprise unless they base their land-use decisions on sound production data. The assistance of professional wildlife managers and biologists may be critical in the planning stages. These professionals can help with planning habitat improvement practices and with determining cost/benefit ratios to help the landowner feel more certain about potential revenues.

Another area of uncertainty is concern about liability for the actions of one's future customers. Fear that one may be unable to control the actions of others, and that this may have legal ramifications, may be enough to dissuade landowners from considering granting access, whether for a fee or at no cost.

Landowners are also legitimately concerned when public policies and government regulations are changing rapidly or unpredictably. Before embarking on wildlife management enterprises that may require several years before there is a return on the investment, they want to be certain that the regulations governing them will not change. Ranchers in Colorado found their hunting enterprises endangered when the Colorado Division of Wildlife decided, in the late spring of 1971, to change drastically the big-game hunting season in order to reduce the number of nonresident hunters and to let the population of big game increase (Shelton 1978). Instead of the traditional three-week season for harvesting both deer and elk, the Colorado Division of Wildlife split the deer and elk seasons and prohibited any big-game hunting for five days between the two seasons. The division also reduced the bag limit on deer and elk. The results were dramatic. The number of nonresident hunters dropped 45 percent, which resulted in a loss of income to the division of more than $2 million (Feltner 1972). There were 89 percent fewer nonresident sportsmen licenses and 53 percent fewer nonresident deer licenses sold. Three cooperating ranchers reported that their reservations were down by more than 50 percent.

The potential effect on ranchers' incomes from a sudden change of season is obvious. Not so obvious are the effects of lost opportunities. Just after these season changes occurred, a certified public accountant in Grand Junction, Colorado, reported to Shelton (1978) that a client of his had been offered $25,000 by a nonresident for the hunting rights on the client's land for a period of ten years. The amount was to be paid when the lease was signed. After the season changes were announced, the potential lessee feared that nonresident hunting might be prohibited in the future, or perhaps limited to specific areas. The nonresident wanted to add a clause to the contract stipulating that there would be a pro rata reimbursement of funds if, by reason of actions by the Colorado Division of Wildlife, he could not hunt on the ranch. The landowner would not agree to the clause and lost the $25,000 in income. Some of this rancher's lands, prime winter range for mule deer, are now broken up into miniranch residential developments. On a national scale, lost opportunity costs to landowners caused by similar regulatory uncertainty could be substantial.

Weak Trespass Laws

Property rights refer to the control one has over owned resources. Indefinite property rights often lead to the depletion of wildlife. Where wildlife resources are concerned, property rights of individuals are not well defined by law.

Remember that in the United States wildlife are owned by everyone and held in trust for the people, while most of the land is owned by individuals. Currently the primary way landowners can claim greater rights to the wildlife on their lands than others have is through the trespass law. Trespass laws to protect property rights vary among states and often are inconsistent. If access cannot be controlled, both landowners and lessees may be reluctant to invest in wildlife enhancements because they are not sure they will reap the rewards of their efforts.

Poaching and trespassing are serious deterrents to wildlife development on private lands. Laycock (1981) reports that two million deer are killed illegally every year, and poaching may be on the rise (Musgrave, Parker, and Wolok 1993). In a survey of forest landowners in the South, Shelton (1969) found that the greatest constraint on the development of wildlife was lack of trespass control.

Landowners cannot be expected to make investments for wildlife enhancement if they cannot reasonably expect to harvest the benefits, either through personal enjoyment or income gained from recreational enterprises. No landowner will invest in a crop of corn, soybeans, or wheat if there is great uncertainty about whether he, his neighbor, or someone else will put a combine in the field to harvest the crop.

Long-Term vs. Short-Term Leases

Some landowners are reluctant to lease their property for hunting or other recreation because they fear that lessees will interfere with their timber or agricultural operations, damage property, or, in some cases, secure easements for developments such as utilities that may prove difficult to remove later. Most leases in the United States are on an annual basis, perhaps because landowners want to "wait and see" how the situation turns out or because they hesitate to commit to a long-term arrangement that may prove undesirable (see appendix B). Often, however, it is the client desiring a long-term lease who may contribute most in the way of habitat enhancements and bring a certain stability to the landowner-client relationship.

In the early stages of a leasing program annual leases can be justified, but after a cooperative landowner-client relationship has been established, longer leases may benefit both parties. Many of the older hunting clubs in the South invest substantial sums of money in road building, equipment, fences, fire lanes, bridges, law enforcement, and wildlife enhancement projects that benefit the landowner as well as the hunting group. These investments will not be made if the hunters are not assured they will benefit from their expenditures. A number of hunting clubs in Mississippi are active in deer management. This usually involves an in-creased harvest of antlerless animals, a reduced harvest of bucks, and habitat improvements—all of which improve deer herd health and productivity. How-ever, these programs may require three to five years to produce results. If sports-men are uncertain about how long they will have a lease, they might not attempt to manage for quality deer.

The preponderance of annual hunting leases may constrain the development of wildlife on private lands by creating uncertainty among sportsmen and inhibit-ing their investment in land improvements and by masking the potential benefits of long-term leases. The truth is, long-term leases may be just as beneficial to land-owners, who need long-term arrangements with the fish and wildlife agencies if they are to take an interest in sustainable use of the wildlife on their properties.

Financing Wildlife Enterprises

Banks and other lending institutions are often hesitant to make loans for the development of wildlife enterprises. There are two main reasons for this: 1) a shortage of economic data on which to evaluate potential returns on recreational businesses; and 2) the fear that in a recessionary period recreational businesses might not be profitable. If such loans are made, they are often at higher interest rates than would be charged for other purposes. Even after the wildlife enter-prise has a proven track record and some stability, those interest rates might not be lowered.

For a long time the Farm Services Agency would not make loans that were strictly for recreational development. Loans for recreational enterprises had to be supplemental to other farm income.

The methods financial officers use to appraise assets may be outdated in the field of wildlife recreational enterprises. Loan personnel are usually eager to learn about new enterprises, but we have failed as wildlife professionals to collect and present data that would enable them to more realistically evaluate the potential of wildlife recreation.

Taxation

Landowners who wish to manage the wildlife on their lands may find stumbling blocks in various tax laws. The property tax in general may discourage wildlife conservation. If landowners develop their wildlife resources to produce income, will they raise their tax base and incur higher property taxes on developments that may not return the expected revenue? In most states agricultural enterprises qualify for lower property taxes. But also in most states wildlife recreational enterprises are not defined as agriculture. (Some wildlife ranchers keep herds of cattle solely for the purpose of qualifying for the agricultural property tax rate.) Texas became an exception in 1991 when the legislature amended the tax code to include raising or keeping wildlife under the definition of agriculture and included use of land for wildlife management in the definition as well.

The federal tax code also poses problems. According to an opinion offered by the review staff of the Jackson, Mississippi, District of the Internal Revenue Service, landowners who manage their land solely to increase the production of wildlife and who sell access permits, memberships, and leases are not considered agricultural landowners for federal income tax purposes. If this opinion is correct, they cannot qualify for certain expense deductions allowed to landowners. This opinion has little effect on the development of wildlife enterprises at present because most individuals who are developing such projects already qualify as agricultural landowners for tax purposes. However, this could be a serious constraint in the future for those who might seek revenue solely from wildlife enhancement.

Estate and inheritance taxes are very real constraints to the conservation of wildlife habitat. Too often heirs must sell rural lands to pay these taxes; most often the sales are to developers who alter the land to the detriment of wildlife.

Marketing Arrangements

The absence of an orderly marketing system is conspicuous in the field of wildlife recreation (Davis 1964). This can be an especially daunting constraint for farmers and ranchers unaccustomed to seeking out and developing markets on their own. The many marketing outlets and the infrastructure that guide them in marketing other agricultural products simply do not exist at present for wildlife recreation. Yet there is no lack of willing customers, and the demand for this "product" is continuously increasing. What is needed is a system that can bring these buyers and sellers together in an orderly and mutually beneficial way.

Table 6.1 summarizes some of the constraints to wildlife stewardship on private lands.

While these constraints to wildlife management by private landowners are significant, all can be overcome. It is up to governments, recreationists, and landowners to find ways of working together constructively with trust and goodwill, keeping in mind that their common goal is the preservation of our wildlife resources.

Table 6.1.

Factors that Encourage and Discourage Wildlife Conservation on Private Lands

Encourage	Discourage
Financial incentives (cost/share)	No financial incentives through state/federal cost/share programs
Agricultural classification for income tax purposes (Schedule F)	Nonagricultural classification for income tax purposes (Schedule C)
Agricultural classification for property tax	Nonagricultural classification for property tax
Property tax credits for conservation	No property tax credits for conservation
Investment tax credits for conservation	No investment tax credits for conservation
Inheritance/estate tax relief	Current high inheritance/estate taxes
Strong trespass laws	Weak trespass laws
Liability relief	Liability exposure
Long-term leases and federal/state partnering agreements	Short-term leases and federal/state partnering agreements
Realistic financial appraisals of wildlife resources	No financial recognition of wildlife values
Lending institutions informed about wildlife enterprises and resources	Lending institutions uninformed about wildlife enterprises and resources

Although a public resource, most wildlife live on private lands.
Photo by Stephen Kirkpatrick.

White-tailed deer and other wildlife consume agricultural crops, here a stand of milo.
Photo by Stephen Kirkpatrick.

Moose may be unwelcome visitors near homes such as this one in Homer, Alaska.
Photo by Stephen Kirkpatrick.

This white-tailed deer is foraging in a soybean field. Photo by Stephen Kirkpatrick.

Sheep are this game rancher's primary livelihood; wildlife management and hunting provide supplemental income and enjoyment. Photo by Delwin Benson.

Many white rhinoceroses are being reintroduced onto private lands in South Africa. Photo by Delwin Benson.

In the cape region of South Africa, wildlife ranching is a common enterprise.
Photo by Delwin Benson.

Hunters talk with a local law enforcement officer. Photo by Delwin Benson.

Both hunters and landowners benefit from long-term leases and ongoing relationships.
Photo by Delwin Benson.

Hunters prepare for a guided hunt on private land. Photo by Delwin Benson.

Managed logging on private lands can benefit some wildlife species.
Photo by Delwin Benson.

Birdwatchers seek access to both public and private lands. This flock of egrets is on Lake Gassoway, Louisiana. Photo by Stephen Kirkpatrick.

These recreationists are watching wildlife on private land near Aspen, Colorado. Photo by Delwin Benson.

Bald eagles and other nongame or endangered species have commercial value for nonconsumptive recreation. Photo by Stephen Kirkpatrick.

This artificial stream was created by the landowner for endangered fish management.
Photo by Delwin Benson.

Wildlife photography is a popular family activity. Photo by Stephen Kirkpatrick.

Fishing on private lands can bring in supplemental income for landowners.
Photo by Delwin Benson.

This herd of bighorn sheep introduced on private lands will attract recreationists.
Photo by Delwin Benson.

Students learn about Native American art at a private ranch. Photo by Delwin Benson.

The creation of wildlife and recreation districts will help in resource management.
Photo by Stephen Kirkpatrick.

Pheasant hunters survey likely habitat on private land in Kansas.
Photo by Delwin Benson.

Campers in Denali National Park, Alaska, are typical of the growing number of recreationists seeking wilderness experiences. Photo by Stephen Kirkpatrick.

There is a role for everyone in wildlife stewardship. Photo by Stephen Kirkpatrick.

CHAPTER 7

Overcoming Constraints

A Plan for Action

Those who stand to gain from the enfranchisement of private landowners for wildlife stewardship ought to share the obligation of making such a policy successful. In short, that means both society as a whole, as represented by its various governments, and the landowners themselves. There are certain actions both parties can take to bring about better wildlife conservation. Recreationists too have certain responsibilities. When we look at the critical relationship that exists among these groups, we begin to understand that their common goals and interests can create a synergy for right actions.

What Governments Can Do

Estate and Inheritance Taxes
In the area of taxation, there is much governments can do to create incentives for wildlife stewardship. Current estate and inheritance taxes can be almost confiscatory in nature. Too often we hear of rural lands that have been held in families for generations but must be sold in order to pay inheritance taxes. The highest estate tax rate is now near 55 percent, while the highest income tax rate is far lower. This is probably the primary cause of land division. Many ranch and farm families are not able to pass on the results of their hard work to family members because of this high tax. We need to look for ways to reduce inheritance and estate taxes, particularly for those who would agree to leave their lands in green space. This single move might do more for wildlife habitat in America than any other factor.

Conservation Easements
Most conservationists would agree that it is in society's best interest for ranches and farms to remain intact. Conservation easements can help save estate and inheritance tax dollars by lowering the appraisal at the time the property is passed

on. The property is appraised for the highest and best economic use at the time and then appraised again based on its remaining in its agricultural or natural state. The difference between the two appraisals can be deducted on income tax returns in the year of the gift and, if need be, carried over for up to five years. The lower appraisal for conservation purposes reduces the taxes ultimately due when the property is passed on to family heirs.

To improve the acceptance of conservation easements, we need to find a way to help property owners cover the up-front legal and appraisal costs. Many landowners who would consider putting their land in conservation easements believe they are doing enough by giving up their potential incomes. They do not want to pay the $15,000 to $40,000 required to produce the legal documents and pay lawyers' and accountants' fees. It might be feasible to form conservation foundations that would cover these costs, thereby preserving large acreages at a fraction of what the purchase cost would be. We also need more entities willing to hold these easements. The law requires that such entities be state or federal agencies or nonprofit wildlife or environmental organizations. Because it can be costly for private organizations to monitor the easements they hold, we need to find ways of overcoming this burden.

Property Taxes

Property taxes generally discourage conservation, whereas they could provide incentives for maintaining open spaces, wildlife habitat, and animal populations (Shelton 1982). Property taxes based on the highest economic use of the land promote development. In most states there is a great difference between the property tax one must pay if land is appraised according to its highest economic use and the tax required if it is appraised as agricultural land. Often the difference is ten times greater or more. Property used for green space or conservation and not used for agriculture is generally taxed at the highest, development, rate. At the very least, wildlife lands should be classified as agricultural lands. Or perhaps a new "wildlife" designation is needed which would carry an even lower property tax rate. Lands used for hunting, wildlife viewing, and general recreation tend to be large, undeveloped areas or areas developed to improve their natural attributes. They do not require special services provided by cities, counties, or states. When landowners implement conservation practices, such as deferring or eliminating cattle grazing, in order to benefit wildlife, they should not be penalized by paying higher taxes. However, excluding cattle grazing cannot be a requisite for wildlife land status because grazing might be necessary to manage some wildlife, including endangered species.

Minnesota has initiated an innovative property tax credit to preserve wetlands and native prairies. Land preserved as green space is taxed at a lower rate; landowners actually receive a tax credit to offset other taxes if they preserve wetlands and native prairies.

Texas has a helpful property tax code that allows land to be classified as open space used for "wildlife management." To qualify, the landowner must actively manage for wildlife, and at the time wildlife management was begun the land must have been appraised as agricultural land. Another requirement is that the landowner must employ three of seven designated practices in order to propagate a sustained breeding, migrating, or wintering population of indigenous wild animals for human use, including food, medicine, or recreation. These seven wildlife management practices include:

1. controlling habitat;
2. controlling erosion;
3. controlling predators;
4. providing supplemental supplies of water;
5. providing supplemental supplies of food;
6. providing shelters; and
7. making census counts to determine animal populations.

The state comptroller's office, with the assistance of the Texas Parks and Wildlife Department and the Texas Agricultural Extension Service, provides guidelines to each appraisal district to help the chief appraiser determine whether land qualifies for this classification. Different guidelines are provided for each of the ten ecological regions of Texas. These guidelines contain a matrix of practices suggested for that region and lists of the species that are affected by them. The appraiser also requires landowners to provide management plans indicating the time required for certain land and wildlife management practices. Needless to say, there were certain political considerations in the passage of this tax legislation. Counties depend on property taxes to finance their operations, so the legislation had to be revenue-neutral in order to pass. The term *human use* was broadly defined to include all uses, both consumptive and nonconsumptive. This encouraged a broad constituency of interest groups to support the legislation.

Federal Income Taxes

As far as the federal income tax is concerned, landowners managing wildlife properties for income should be classified as agriculturists (using Schedule F). More tax deductions should be allowed for wildlife enhancements, whether for the purposes of aesthetics or increased income. It has long been a feature of our federal tax code that those behaviors beneficial to society are encouraged while those detrimental to society are discouraged. Income taxes are generally thought to be neutral where conservation is concerned. However, there are several provisions in the income tax code that can significantly affect the taxes incurred by landowners interested in producing wildlife.

If the previously mentioned opinion of the Jackson, Mississippi, staff of the Internal Revenue Service is correct (that landowners who manage their lands solely to produce wildlife and who either sell permits or memberships or lease their property for profit are not considered agriculturists for tax purposes), then wildlife managers who hope to profit from conservation practices cannot qualify for certain generous expense deductions allowed to farmers. If farmers who combine agricultural production with wildlife production for profit may deduct expenditures for practices such as clearing land, eradicating trees, treating or moving earth, diverting streams and water courses, building terraces or levees, or constructing water or sediment detention dams, those who manage solely for wildlife production should enjoy the same tax benefits. These practices may be just as beneficial to wildlife as they are to agriculture. In fact, some may serve dual purposes by creating or improving fish ponds or providing water for waterfowl habitat. Obviously the tax code must be reviewed. Farming for wildlife should be on the same footing as farming for corn, wheat, or timber products.

Investment Tax Credits

Tax credits have been given to encourage investments that either stimulate the economy and/or assist in producing products, goods, or services that society desires. Some investments are, by their nature, more risky than others and involve either long-term planning or long time periods to produce results. Timber production is an example.

Before legislation, beginning in 1986, removed most of the tax credits, farmers or ranchers not only could deduct the ordinary expenses incurred in the operation of a supplemental wildlife business but also could take advantage of a 10 percent investment tax credit if they acquired new or used depreciable property with a useful life of at least four years. Examples of implements that aid in the production or harvesting of wildlife and that met this requirement are decoys, guns, siphons, water pumps, boats, tractors, pond control devices, and controlled burning equipment. Landowners could also build structures such as tree stands and duck blinds, assign them the cost of labor and material, and apply these costs toward an investment tax credit. Perhaps some portion or variation of this should be reinstated to favor the production of wildlife on privat.

On October 14, 1980, the president signed into law two tax incentiv. virtually all persons who plant trees on their property. The maximum expenditure eligible for this tax treatment is $10,000 per year. It works like this. If a landowner spends $10,000 to plant trees (for such things as site preparation, seed or seedlings, and labor), he can subtract a 10 percent investment tax credit ($1,000) from the amount of income tax he otherwise owes. The owner also can deduct from yearly earnings the full $10,000 over a seven-year period (in general, $1,428 per year). This new tax incentive could benefit wildlife, especially if it were expanded to include such things as the construction of greentree reservoirs (timbered areas flooded in the winter for waterfowl), fish ponds, watering devices, or other wildlife enhancement projects.

The mechanisms for such a program are already in place. The Farm Services Agency (FSA) oversees cost-share programs for wildlife preservation. This national program could designate wildlife enhancement projects that qualify for the tax credit. The Natural Resource Conservation Service (NRCS) is already required to inspect and list specifications for wildlife enhancement projects. Landowners whose proposed projects were approved by the FSA and inspected upon completion by the NRCS could be given a document to be filed with their income tax returns, allowing for both tax credits and the depreciation of the projects. Another possibility might be to make wildlife enhancements tax deductible on lands not held for income, if approved by these two agencies. Thus, landowners who hold property primarily for aesthetic or recreational reasons could invest in habitat improvement practices to benefit wildlife and receive a reasonable deduction on income taxes owed.

Liability

Even in this litigious age the threat of liability exposure from granting recreational access to one's land is more imagined than real. However, many landowners perceive that there are significant liability risks, so the issue must be dealt with. The American Forest Products Industries, Inc. developed a model liability relief law that has been adopted by many states. The law, however, primarily gives liability relief only to those who do not charge for recreational use. States might find it advisable to revisit this issue. Liability should be reduced for landowners who allow low-density, nonintensive, wildlife-based recreation on private lands whether or not they do so for monetary gain and especially if recreational income is not the primary source of income from the land.

Texas law limits the liability of landowners for injuries or damages suffered by trespassers or those who are invited to use their land. The liability of an owner, lessee, or occupant of the land is limited to a maximum amount of $500,000 and $1 million for each single occurrence of bodily injury or death, and $100,000 for each single occurrence of damage or destruction of property if the injury, death, or damage is caused by an action or omission of the landowner. This monetary stipulation applies only to landowners who have liability insurance coverage in the amounts described. Landowners who do not charge for entry to their lands or whose total income from such charges does not exceed twice their ad valorem tax for the previous year are protected from liability to the extent of the duty owed to a trespasser. These known liability limits at least provide some certainty that if an accident did occur, the landowner's land and livelihood would not be jeopardized.

Exposure to liability also can be reduced by legislative action. About twenty-one state legislatures have given landowners limited liability protection for hunting accidents that occur on private land. Normally landowners who charge access fees are not covered, although Maine and Michigan have exempted landowners from liability even if they do charge access fees. Colorado has a law limiting the liability of professional equestrians and another law defining the responsibility of skiers and ski resorts. In any event, states can only exempt landowners in cases of mere negligence, which means that the landowners are exercising reasonably prudent care to protect those who use their lands. Gross negligence or utter disregard for the physical safety of others may still be subject to liability.

Landowner-Friendly Regulations

Another way governments can help overcome the constraints to private wildlife management is to make sure regulations concerning hunting permits, harvest quotas, and season length and timing make it possible for landowners to benefit fully from the wildlife they are producing. Landowners should be given some number of hunting permits to distribute or use as they see fit. The number would be based upon an allocation system commensurate with the habitat the landowners are preserving and the animals they are producing. Harvest quotas must be established on an incentive basis so that landowners who produce more and better quality wildlife may take more of the biological surpluses if they desire. Ownership of wildlife is not a requisite to this recommendation; in fact, unwarranted debates about wildlife ownership create more controversies than solutions. The concept of enfranchisement allows governments to empower landowners to make decisions about their use of wildlife resources within the context of their management plans. Finally, the length and timing of hunting seasons on private lands must coincide with the needs of wildlife, hunters, and landowners.

What Landowners Can Do

Landowners who embark upon wildlife-based recreational enterprises also have certain responsibilities. First, they must evaluate their land, wildlife, and human resources to determine whether such a plan is feasible (see appendix A). In doing so, they should be honest with themselves about the strength of their commitment to the investments that will be required. These investments include labor, money, and, most likely, time spent gaining new knowledge about habitat and wildlife management practices.

The development of the wildlife management plan is crucial and should be done in cooperation with and with the assistance of professional wildlife managers and state agency personnel. Together they should determine existing wildlife populations and establish which management practices will be most beneficial. Plans should include provisions for managing both hunted and nonhunted animals, but management practices may not be needed (or affordable) for all species. For example, it may be that no new management practices are needed for hunted species. Indeed, the way a farm or ranch is managed may be the reason that these species are present to begin with. However, if nonhunted wildlife on the property would benefit from management, then agreements could be made for their management while enfranchising the landowner to operate a hunting enterprise.

To be successful, landowners must practice good business management and use creative leasing strategies (see appendix B). Long-term leases offer benefits both for the landowner and for wildlife, and this is what the landowner should strive for. For example, rather than charging $4,000 annually for a five-year hunting lease, the landowner could charge an up-front payment of $20,000 (assuming it is not prohibitive for the potential lessee). Invested at 12 percent interest, the up-front payment yields $10,000 more in future income than the annual payments would yield ($35,236 vs. $25,411). This makes possible a greater investment in developing and sustaining the wildlife habitat and population. Long-term relationships between landowners and recreationists also foster a sense of shared stewardship and make it much more likely that the recreationists also will invest in enhancements to the land.

Landowners must communicate clearly with their customers about rules of behavior and provisions in the contract. The best time for differences of opinion to occur is in the planning stage before either party has an obligation to the other. Open discussion about all aspects of the lease gives the landowner an opportunity to evaluate the potential customer's character and philosophy in a businesslike manner without being offensive. If land use and the customer's behavior cannot be agreed upon at the beginning, then it is best that no contract be entered into. The discussion should include what the landowner expects and a complete explanation of the risks customers may face in using the property. Communication should be frequent and open throughout the term of the lease.

Most landowners currently operating recreational enterprises reduce their liability exposure by shifting the risk to insurance companies, by incorporating the wildlife enterprise separately from other operations, or by informing recreationists of all known hazards and requiring them to sign liability-release statements. These risk management techniques can be very helpful.

Usually the larger the area under management, the more effective that management will be. Landowners should consider forming cooperatives, conservancies, or associations with adjoining landowners. Decisions landowners make about the use of their property may affect their neighbors. Wildlife are more apt to thrive when their habitat is managed on a large scale and is not fragmented.

Landowners who begin recreational enterprises should also pay attention to customer satisfaction. What improvements or amenities might they offer to make the recreational experience more enjoyable? On lands leased for hunting such amenities might include supplemental feeding plots or blinds. On other lands they might include the construction of walking trails, shelters, or campgrounds. Much depends on the extent to which landowners wish to invest labor and money in such improvements and the willingness of customers to pay more for them. In any case, landowners should consider alternatives and have a basic business plan to follow.

What Recreationists Can Do

Those who desire greater recreational access to wildlife should make their desires known both to landowners and to their state governments. Many landowners are unaware of the recreational potential of their lands and might be encouraged to grant access if they understood the demand. Likewise, state parks and wildlife agencies might be encouraged to promote recreation on private lands through cooperative agreements with landowners. However, the granting of access carries certain responsibilities. Recreationists should be willing to pay for access opportunities, and they must honor such privileges with ethical behavior that in no way interferes with the landowner's other business operations and lifestyle.

What has been described here is a relationship among government, landowner, and recreationist based upon trust and mutual respect. That is the essence of landowner enfranchisement.

CHAPTER 8

The Shape of the Franchise Agreement

We have been discussing constraints to the development of wildlife and recreation on private lands. Franchising is a logical approach to removing or reducing those constraints. Franchising allows the state to join with private landowners in sustainable wildlife enterprises. The logic of franchising lies in the state's sharing a valuable set of rights with a private producer. The state agency doing the franchising will neither lose control of any valuable rights nor pay any subsidies; it gives up none of its public trust or responsibility nor its ability to issue licenses and enforce its regulations. The agency merely designates certain private operators to deliver services that are increasingly difficult for the state to produce—namely, habitat improvement and high-quality recreational opportunities. In doing so, it expands the resources available for wildlife management and conservation.

The security of a franchise revolves around the contract between state agency and landowner, and relies on the commitment of each party to perform certain duties under the contract. In Colorado's Ranching for Wildlife program each ranch manager under contract agrees to implement an approved habitat management plan and to provide fee hunting for a period determined with the state. The ranches provide access to free hunting to members of the public selected at random from a pool of applicants. The Colorado program allows the state to perform its responsibilities for the conservation of wildlife, the ranchers to receive income from hunting, and the state economy to benefit from an infusion of expenditures by nonresident hunters, who tend to be the clients of the ranches. In this way franchising may reflect a natural trend toward more private responsibility for wildlife management (Hudson 1993).

The private landowner can expect several things from the contract with the state agency. First, the contract affirms the state's responsibility for the conservation of wildlife but conveys to the landowner the rights to extended hunting season length and timing and a claim on a share of the wildlife to be harvested from the property.

The Wildlife Management Plan

Second, the contract requires that a wildlife management plan be developed jointly by the landowner and the agency. The plan commits the landowner to certain performance and specifies the actions that will be taken to make the franchise successful.

The plan begins with a summary of the landowner's long- and short-term objectives for consumptive and/or nonconsumptive uses of wildlife and other recreational activities. It briefly describes current land and management practices that may affect wildlife and recreation, such as livestock grazing, timber production, energy development, and farming. Plans for changes in normal agricultural operations should also be included.

The Wildlife Management Plan contains a legal description of the property included in the franchise agreement, including all lands to which the landowner holds the deeds and all public lands (Forest Service, Bureau of Land Management, state school lands, etc.) that lie within the boundaries of the privately owned lands. Maps that show the locations and boundaries of private and public lands should be appended to the plan.

The status of each wildlife species to be harvested or affected by land use or management changes is described. The species status report includes the number of animals on the property at various times of the year; the number of post-hunting-season young (per one hundred adult females); the number of post-hunting-season males (per one hundred females); and the number, age, and sex of animals harvested during each hunting season. For a nonhunting enterprise the abundance and location of nongame species are important information for understanding long-term trends, for aiding recreationists, for promoting recreation on the property, and for monitoring the effects of recreation to make sure it is not detrimental to wildlife or wildlife habitat.

The general status of the habitat for each species is included. This can be presented as a list of habitat or vegetative types, along with the percentage of the total land on which each type is found. There should also be a description of the general condition and use of each type of habitat and vegetation. Vegetative maps may be available from the Natural Resources Conservation Service and are valuable in determining habitat status.

At the heart of the plan are the landowner's management objectives, listed by species and areas of the property. These must be measurable objectives (where, when, how many, etc.). Examples of objectives for a game species might include, but are not limited to:

1. increasing or decreasing the population of a species by a specific percentage in a certain period of time;
2. maintaining specific ratios of young animals (e.g., per one hundred adult females and males, per one hundred females and adult males, per one hundred females, etc.);
3. harvesting a specific percentage of the population by sex or age class each year;
4. manipulating habitat by a specific technique to obtain an expected result; and
5. making special provisions for protecting nonhunted threatened or endangered species.

Examples of objectives for nonhunted species could be:

1. maintaining a valuable habitat component;
2. involving recreationists in land and wildlife management activities; and
3. lessening the negative impacts of human presence and maximizing the benefits.

General management objectives are also stated. These would include the kind of recreational enterprise to be operated and who the expected customers will be.

Next in the plan are the management strategies and methods that the landowner will implement to meet the objectives. They may be listed by species, by ecosystem, or by clientele group. Examples of management strategies might be:

1. plans for improving habitat, including a map of locations;
2. preferred hunting dates and/or bag limits to meet herd objectives and satisfy customers (showing the relationship of the objectives to existing season length and timing, permit availability, and other governmental regulations);
3. plans for enhancing the habitat of endangered species; and
4. amenities to be provided for recreationists.

Finally, the management plan specifies how the landowner will evaluate success in reaching management objectives. An evaluation should be conducted annually and might include measuring:

1. species population counts and trends over time;
2. the number of recreationists using the land;
3. the total number of animals harvested by species, age, condition of animal, measurements, location, date, etc.; and
4. the number and kind of habitat improvement practices implemented (number of acres burned, cleared, seeded, etc.).

Landowners may be surprised to learn that recreationists are often eager to assist with data gathering. For example, birders often participate in national counts of wintering and breeding birds. "Average" citizens such as these may willingly help with management objectives ranging from population estimation to habitat improvement.

Delineation of Responsibility

The third aspect of the franchise agreement is that it clearly delineates the authority and responsibilities of the state agency and the landowner. In the simplest terms, landowners have the responsibility to carry out sound wildlife, habitat, and human management plans, while the state has the responsibility to assist landowners and hold them to their commitments. The difference between a true franchise arrangement and most leasing systems that currently exist is that a lease involves one person leasing a privilege that is not used or shared by another, whereas a franchise exists between two parties who share a common goal and who work together to achieve that goal. In a franchise agreement neither party may make unilateral decisions that affect the integrity of the agreement or that jeopardize the stated goal.

For example, both parties have roles to play in determining wildlife populations. Generally landowners will supply species counts on their properties (and may hire professional biologists to help them do so), while state wildlife management professionals will compile general data gathered on larger scales from the state's management units. The state agency has the authority to determine the length and timing of the hunting season; but rather than doing so on a statewide basis, it should consider regional game species populations and the needs of landowners within each region. The management objectives of franchisees within a region are important considerations. Hunting-season length and timing must allow optimal enjoyment for hunters and provide landowners with a reasonable return from providing access to wildlife.

The harvest quotas that will be in effect on a property should be agreed to jointly by the landowner and state agency and should be based on population surveys and management objectives. Landowners who produce more wildlife must be rewarded with larger harvest quotas. There are two ways in which the agency can give landowners permits for harvesting their properties' quotas of game animals. Landowners might be required to pay a fee for each permit, which they would recoup by leasing their lands for hunting. Or the agency might not charge for some of the permits, in exchange for a period of free public hunting on the property. When decisions such as these are made jointly, landowners have an incentive to become the state's partners in conservation.

The Length of the Franchise

Finally, the franchise agreement stipulates the period of time for which the agreement will be in effect. This should be long enough to reduce the uncertainty of all parties, encourage landowners to invest in environmental improvements, and give landowners the opportunity to profit from their efforts. Ten years seems a minimum amount of time for such an agreement, and the franchise should be renewable well in advance of the tenth year so that the landowner can plan into the future. The landowner who is living up to the contract should expect to have a perpetual franchise.

Monitoring Franchise Agreements

Part of the government's responsibility is to monitor the success of wildlife management franchises, but care should be given to the administrative burden imposed. The temptation to exert too much control by micromanaging each property or by making arbitrary decisions can create problems. A laissez-faire approach would not be advisable either. We can safely assume that landowners who would want to participate would be motivated to exercise good land and wildlife stewardship and that this desire would be reflected in their plans. So state monitoring can be most beneficial if it takes the form of frequent communication with landowners and the offering of technical assistance. The state also can ask to see landowners' annual evaluations of their management plans, including the status of plants, animals, and the recreational enterprises. Franchise agreements are important because through them governments can empower landowners to make decisions about using wildlife with some security and certainty about the future, as long as they meet their obligations for good stewardship. Of course, the state must meet its obligations as well.

CHAPTER 9

Shared Management of Nongame and Endangered Species

Hunters and sportsmen's organizations recognized early on the value of private lands to wildlife conservation, and they contracted with private landowners to gain access to wildlife for recreation on those lands. Consequently, hunting offers history on which to base recommendations for franchise arrangements for wildlife management. However, there is every reason to believe that these same arrangements may benefit the future management of nongame and even endangered species, providing new nonconsumptive recreational opportunities. A challenge of the next century is how to make nongame and endangered species valued and managed components of private lands.

We must not gloss over the situation that presently exists, however. The passage of the Endangered Species Act in 1969 altered the relationship between private landowners and the government in ways that have too often forced them into adversarial roles. At times the chasm between them has seemed so wide that it takes tremendous effort to bridge it. But as responsible citizens we all must make that effort. No purpose is served, certainly not sound wildlife management, if we continue on the present course of animosity and conflict. To understand the difficulties that exist, we must understand exactly what the Endangered Species Act is and what it allows the federal government to do.

The Endangered Species Act is a far-reaching and powerful environmental law that affects us all, especially those who own property. The act provides for the listing of threatened and endangered species and the protection of those species on both government-owned and private lands. The law mandates that listed species cannot be harmed in any way, and the alteration of their habitat is forbidden. The very nature of the law has made it a threat to private property rights in the United States, because when a threatened or endangered species is found on private land the owner of the land may be prohibited from taking any action that could alter the population of the species in any way. Landowners may not even be allowed to continue the operations from which they make a living. The Fifth Amendment to the Constitution states that landowners are to be compensated when their lands are taken by governments for public use. The government argues that enforcing the Endangered Species Act does not amount to "taking" land, but landowners could argue otherwise.

Most Americans support the protection of threatened and endangered species as the natural and logical thing to do. Yet most Americans probably feel that the issue has little impact on them. As our society becomes increasingly urban, most Americans have little if any direct contact with wildlife other than those animals that are well adapted to urban life. The truth is that all Americans are in one way or another affected by endangered species. We pay the bills for government actions, and when those actions increase the cost of commerce, we pick up the tab again as consumers (National Wildlife Institute, 1994). Therefore, it is important for all of us to understand the issues.

Few informed people would disagree with the idea that preserving the biological diversity of our world is a worthy goal. Protecting threatened and endangered species is a way to work toward this goal for the good of society. In its present framework, however, the law which seeks to protect plant and animal species may do more harm than good. It is easy to be confused by the sensational stories that so often appear in the media and make it seem that we must choose up sides. Do we support the protection of plants and animals that are in danger of becoming extinct? Or do we sympathize with the people whose lives will be adversely affected by government efforts to save those species? The sad result of such polarized thinking is that it becomes harder and harder to see the middle way. It is unfortunate that so many people believe only rigid government control can preserve threatened and endangered species and ensure biological diversity. Landowners are often viewed as the enemy by environmentalists who assume landowners care little for any species that does not yield a profit. On the other hand, landowners see that environmental laws are often made by people who are disconnected from the land, and they fear that under the influence of such people government will try to take control of their land. Are these perceptions accurate? Are we so deep into our respective trenches that we can no longer see common ground or creative solutions?

Effects of the Law on Private Property

The burden of enforcing the Endangered Species Act falls, for the most part, on the U.S. Fish and Wildlife Service for nonmarine species. This organization is chronically understaffed and underfunded, and is thus unable to offer meaningful assistance to help landowners understand and comply with the law. The unfortunate result is that the service's approach is too often to tell landowners what they cannot do, without helping them learn what they can do. This situation understandably causes landowners to fear that government will take control of their lands and livelihoods. In some cases that is exactly what has happened. A few examples will suffice.

A North Carolina timberland owner consistently tried to harvest trees in a way that provided habitat for wildlife. Campers, hunters, and fishermen often used his land because he believed that outdoor recreation and tree farming were compatible. But when the red-cockaded woodpecker arrived on his property, the Endangered Species Act put one thousand acres of his land off-limits to him, which caused a loss of about $1.8 million in timber sales. To protect his remaining land from being occupied by the bird and consequently falling under federal control, he harvested trees before they became old enough to attract the woodpeckers (National Wilderness Institute 1994).

A landowner in Utah planned to build a golf course and campground on his property. Neighbors had long used the spot for recreation. The project was halted, however, when the U.S. Fish and Wildlife Service declared that a pond on the property was habitat for the endangered Kanab amber snail. The area was fenced off, people were no longer permitted on the pond's banks, and the landowner was forbidden to work in the area. He was never compensated for his losses, estimated at $2.5 million (National Wilderness Institute 1994).

In California a Chinese immigrant bought land on which to grow Chinese vegetables for sale to the Asian community. The land was zoned for farming, and county officials had told him no permit was needed. However, his equipment was confiscated and he faces criminal charges for violating the Endangered Species Act because his tractor allegedly disturbed the habitat of the endangered Tipton kangaroo rat (National Wildlerness Institute 1994).

These stories illustrate the assumption government sometimes makes—that landowners must be forced to preserve endangered species and their habitat because, left to their own devices, they will not be good stewards of the land. Of course, not all landowners manage land resources well. There are too many occurrences of overgrazing by domestic livestock, for example. However, it is a myth that farmers and ranchers today still have the kind of frontier mentality that says "Exploit the land, use it up, and then move on." That way of life ended long ago (Horton 1992). For many, their land is their only means of economic survival, and they must become adept at taking care of it. Those who make their livings from the land, who consider it a privilege to own land, and for whom the land represents not just income but a valued way of life are often the best environmental stewards among us. Their careful husbandry of our natural resources is responsible for the biodiversity we now have and seek to save. Getting rid of false assumptions about how landowners will behave in the absence of strict government control is the first step to building the kind of trust required for meaningful dialogue between governments and landowners. When landowners are respected for their knowledge and care of the land, and when governments have the structure and laws that allow them to work with rather than against landowners, real progress can begin.

Finding the Value of Nongame and Endangered Species

We have shown that when landowners are free to benefit from the commercial value of wildlife, they are far more apt to conserve that wildlife. The same is true even when the wildlife are nongame or even endangered species, because they, too, have commercial value. We must build a bridge between the interests of society (preserving species) and the interests of landowners (maintaining their private property rights and benefiting economically from their property).

Shared management is a logical approach. When a threatened or endangered species is found on a property, government agencies could work with the landowner to develop a management plan agreeable to both parties. The plan might include specific ways to preserve habitat without interfering unnecessarily with normal land operations. For example, the endangered Kirtland's warbler is known to nest only in a small area of north-central Michigan. Its required habitat is thickets of young jack pine trees. To preserve the birds' habitat the forests must be continuously managed to promote the growth of new trees, but the removal of large, old trees does not affect the warbler. Another example is the golden-cheeked warbler, which nests in old-growth cedar in the Texas Hill Country. Landowners there are allowed to remove brushy young growth without disturbing the birds' nesting habitat.

Besides helping landowners manage species without losing agricultural income, agencies might also help landowners determine how to establish wildlife-based recreational enterprises that do not disturb or harm the animals. Possibilities include bed-and-breakfasts, nature trails for birding walks, and guided photo safaris. Ted Eubanks, noted Texas environmentalist, sees great promise for ecotourism as a method of funding conservation. "I have a strong belief in private sector environmental solutions," says Eubanks. "Rather than having the government do everything, how do we get local people involved? I strongly believe profit is a powerful conservation tool" (Acker 1995).

Establishing Just Compensation Systems

Some endangered species do not lend themselves to ecotourism opportunities, and some landowners will not find recreational enterprises compatible with their other land operations. When a landowner must lose income in order to maintain habitat, there should be adequate compensation programs available to replace that income.

Indirect compensation for saving endangered species habitat might be in the form of tax relief. Reducing property and income taxes makes good sense when the value of the property and the landowner's potential income are both reduced by the need to set aside land for endangered or threatened species. Tax incentives should be provided not only for landowners who maintain existing habitat for endangered species, but also for those who create it. Inheritance taxes should be reduced for the heirs of landowners who pass well-managed habitat and restored species to the next generation, assuming the heirs make a legal commitment to maintain the habitat and continue managing it. Inheritance tax reform could be used to ensure that large tracts of habitat do not have to be broken apart but rather are maintained for their wildlife value.

It may also be important for government to directly compensate landowners with payments for loss of income. It is appropriate that landowners who manage endangered species on behalf of society be paid with society's funds. Monetary support for landowners might also come from philanthropic and conservation organizations and from bond sales. In a recent article Ted Eubanks voiced his support of direct incentives to property owners who choose to enhance their land for the benefit of endangered species and other wildlife: "If we can pay crop subsidies, I can't imagine why in the world we can't pay for wildlife. It makes sense to me. The problem is really social. The problem is economic" (Acker 1995).

In a paper about re-authorizing the Endangered Species Act (ESA), Larry McKinney, director of resource protection for the Texas Parks and Wildlife Department, notes: "a means must be found and incorporated into the ESA of assisting rural landowners to comply with the act. Indeed, one of the greatest stumbling blocks has been the lack of a means to assist landowners and to provide incentives for action. . . . incentives to rural landowners serve to encourage use of land management techniques that aid in protecting endangered species. If the ESA does not begin to provide positive incentives to private landowners, then the act will continue to be ineffective in achieving its goals on private lands" (McKinney 1993).

Another way to compensate landowners for protecting endangered or threatened species is to allow them to use common and abundant animals as an economic resource on a sustainable basis. Permits could be granted to participating landowners for harvesting game species in exchange for conserving rare species.

Government agencies are now realizing that landowners need to be rewarded rather than penalized for having endangered species on their properties. The Texas Parks and Wildlife Department has begun a pilot incentive program, funded by the U.S. Fish and Wildlife Service, that will make grants of up to $10,000 each to qualified landowners. The Landowner Incentive Program seeks candidates who will be creative in protecting or increasing rare species on their lands while still engaging in traditional farming and ranching practices. Some ideas that could be funded include restoring native vegetation, adjusting grazing rates, controlling fire ants, putting up nesting boxes for birds, and protecting habitat. Program funds also could be used to pay bonuses to landowners who increase rare species or their habitat as a result of conservation efforts. In announcing the program, Gary Graham of the Texas Parks and Wildlife Department remarked, "Rare species conservation in Texas will succeed or fail on private land, since 97 percent of the land in our state is privately owned. The rural landowners of Texas have the habitat, and most of them are good stewards of natural resources. Our job is to assist with practical, financially viable options for them to conserve the resources in their care" (Associated Press 1997).

There is no need for society to fear the loss of endangered species on private lands. There is no need to accept rigid government control as the only solution. "Anything accomplished through government programs can be done outside them and without so much intrusion and waste. The future of conservation doesn't lie in centralized command and control. Forward looking policies will reward good stewardship, tap the free market to accomplish conservation goals, respect and protect private property, and encourage site and situation specific practices rather than sweeping bureaucratic mandates" (Gordon 1995).

The need to amend the Endangered Species Act is now widely acknowledged, and many within both government and the private sector have offered suggestions for solving its inherent problems while retaining its intent. We know that there are ways to include landowners and understand that the so-called conflict between agricultural landowners and environmentalists is far more perceived than real. We know that working together on environmental issues is a better way than perpetuating old animosities. Now we must find the will.

CHAPTER 10

The Holistic Management Ideal

Up to now the debate about the role of the private sector in wildlife and recreation management has focused upon the rights of society, represented by its various governments, versus the rights of private landowners. Wildlife do not benefit from debates, but they will benefit from a holistic approach to management that enfranchises the private sector to play a key role as partners in preserving society's resources.

Private involvement in wildlife and recreation management is both desirable and necessary. Future debate should address how governments can give private landowners appropriate technical assistance and help them evaluate their management plans. There will likely be no more significant additions to the public land and water base. Private lands are a national resource that we cannot afford to neglect. Those who manage private lands need to be brought into the management and decision-making circle. We must recognize that the wildlife they control by virtue of their land ownership do have commercial value, and that commercial value can give impetus to a more reasoned and comprehensive approach to wildlife conservation. Society will be the winner when this is accomplished.

In 1930 the North American Game Policy Committee, chaired by Aldo Leopold, was the first to look at wildlife management on a national scale. The committee identified three alternatives for addressing the question of wildlife on private lands. It said that the government could buy out private landowners and become the manager of wildlife (i.e., acquire more public lands), cede rights of wildlife ownership to the private sector, or compensate landowners for providing wildlife habitat and recreational opportunities. The committee recommended the third alternative as being most realistic.

In the years since that first game policy was written, it has become clear that the committee's thinking was sound. They recognized that the acquisition of public lands takes place slowly, as money and politics allow, and there are definite limits to the amount of land governments can acquire and manage. Certainly there will never be enough public land to preserve all wildlife species or to satisfy the public's desire for recreation. Likewise, the committee understood that ceding the rights of wildlife ownership directly to landowners breaks with North American tradition and would never be an acceptable solution. However, the compensation programs the committee endorsed have offered only interim solutions with little long-term success. They have generally been too small or short-lived to be successful when compared to the economic and social costs of producing wildlife and recreation. Also, compensation programs have been based largely on the assumption that wildlife are detrimental to landowners and, therefore, that landowners who are willing to tolerate the presence of wildlife, so that hunters and others might enjoy access to them, should be compensated for their trouble. This assumption is a negative one. Perhaps we have not been able to understand, until now, that under the right circumstances wildlife can be an asset rather than a liability for the private landowner.

Three New Alternatives

In recent years heightened tensions over government environmental regulations have led some landowners to demand private wildlife ownership (Leopold's second alternative), while others have proposed a fourth alternative—the removal of the public's animals from private land altogether. These few would like to have all wildlife and recreationists off their lands and out of their land-management decisions.

A fifth possible solution is to take all rights of wildlife management and all control over land access away from landowners and give these controls to federal and state governments. That idea is as foreign to North American tradition as the idea of ceding wildlife ownership to the private sector. It is not likely to be a reasonable alternative elsewhere either. In Europe and many other parts of the world the trend is toward more democratic and capitalistic societies, with more opportunities for the private sector.

Is there another possible policy for wildlife and recreation management on private land? Establishing franchise agreements between state agencies and private landowners is certainly a start. As we have seen, these agreements give landowners a voice in determining what will happen on their lands; they encourage landowners to undertake good stewardship practices and assist them in doing so; they make possible the predictable management of wildlife on a larger scale than government agencies could achieve alone; they help to create recreational opportunities for the public; and, perhaps most important, they bring government personnel and individual landowners into constructive partnerships in which the shared goal is caring for the public's wildlife resources.

Ultimately the benefits of enfranchisement may go far beyond the government-landowner partnership. The extreme points of view we have heard in recent years concerning wildlife and the environment have tended to pull our society apart into various factions. Enfranchisement could be the catalyst for ending such divisiveness because it emphasizes the need for all of us to work together to save our natural resources. Perhaps the result will be integrated partnerships involving not just state wildlife agencies and private landowners but also universities, nongovernmental organizations, businesses, communities, and the consumers of wildlife recreation—each group with its own role to play.

How can we establish such an all-inclusive approach to wildlife and habitat stewardship? We must begin with certain assumptions and recommendations about those assumptions.

First, having wildlife on private land is better than not having wildlife. Consequently, land operators must have incentives to care for and enhance wildlife populations and their habitat.

Second, society has an obligation to promote wildlife, wild lands, and recreational pursuits that value those resources as reminders of the past, for enjoyment in the present, and as a legacy for the future. Private landowners have a societal obligation to practice land and wildlife stewardship as part of their normal operations.

Third, landowners can be enabled by the government, through regulations, policies, and technical assistance, to deal with their own needs and problems within the framework of government's mandate to care for wildlife belonging to all of society.

Fourth, private landowners will not unanimously embrace the idea of managing wildlife and recreation on their properties, whether for personal or public benefit. Consequently, those who create regulations and public policies must be sensitive to the effects they may have on landowners. Landowners must be given complete information about how they can participate in private land stewardship and the opportunities that exist for developing wildlife and recreational enterprises.

Fifth, all of us are influenced by our own personal philosophies concerning the value of wildlife. Where these philosophies create conflict with one another, we must seek ways to resolve our differences. That means listening, seeking common ground, and striving to include all points of view in decisions that are made.

Sixth, those who benefit most from programs that enhance wildlife should be expected to contribute to the success of those programs. The primary beneficiaries in the franchise scenario are the state agencies with responsibility for wildlife management, the landowners who stand to gain supplemental income from recreational operations, and the recreationists who enjoy access to wildlife on private lands. In the largest sense, however, the benefits extend to local business communities and, indeed, to all citizens. In broad and truly integrated partnerships, we all must contribute to the success of wildlife enhancement programs.

For a system of shared costs and benefits to work, we must clearly communicate expectations for each group involved. The recreationists who use private lands will be obligated to pay for that privilege. Hunters and anglers have been the traditional users of private lands, and many of them are accustomed to paying for access. They must now be joined by birders, hikers, photographers, campers, and others in supporting private management of wild lands by their willingness to pay for access. The number of people who enjoy such pursuits is growing rapidly, and if more private lands were available for them to use, they would take advantage of the opportunity. Landowners who benefit from wildlife-related recreation have an obligation to maintain and enhance wildlife populations and habitat. Those who contribute more and add greater value to recreational experiences should realize greater returns. Communities that gain economically from local wildlife recreation also should help compensate those who produce wildlife. This could take several forms. Local sales taxes on goods and services associated with wildlife recreation could be put back into the production and promotion of outdoor resources and activities. Or bond issues could support wildlife and recreation management areas on a local or county basis.

Initiatives such as Teaming with Wildlife (proposed in 1998 and modified in 1999) would place a federal excise tax on wildlife-related goods as a means of generating revenues for managing nongame wildlife, similar to legislation in the 1930s (Pittman-Robertson) and 1950s (Dingle-Johnson and Wallop-Breaux) that generated money to manage game species. Unfortunately, there has not been strong support in the legislature for nongame funding.

Wildlife and Recreation Districts

We begin now to see the possibilities of expanding the concept of individual landowner enfranchisement to something far greater. A system of wildlife and recreation districts could be established throughout each state (Benson 1991a). These districts would be centers for the holistic management of wildlife on both public and private lands. Established according to the goals of government managers, private landowners, and public users of the resources, these districts would be based on a willingness to work and negotiate together for the common good. A systematic gathering of data about these goals, and about the status of wildlife resources, would provide a starting point. From there specific management plans could be developed to meet the needs of the districts.

The concept of districts as administrative units is similar to the existing wildlife management unit or soil and water conservation district concept in North America, except that a broader base of publics would be brought into the management model. Another difference would be that all parties who participate in the district would be expected to share in the costs and benefits of management decisions, rather than just provide input to those decisions. A district could be centered around opportunities or problem areas such as where animal damage harms agricultural economies. It could be established to holistically manage areas where public and private lands intermix. A district could be formed to protect large watershed areas or specific habitat types, or for many other reasons.

To further illustrate the concept, consider the difficulties of managing migratory big game populations in North America. Animals may spend summers and early autumns in national parks and forests and live during winters and springs on private lands where they damage fences and consume livestock forage without providing the landowners with any opportunity for income. On other private lands the animals may be present during autumn hunting seasons. When the situation is perfect hunters, photographers, and campers enjoy the animals throughout the year on both public and private lands. Businesses benefit from consumptive and nonconsumptive tourism, and wildlife agencies are funded to maintain animal populations for all concerned. Needless to say, the conflicts that arise are difficult to address because the costs and benefits the animals represent are not shared. However, if traditional land and wildlife managers took the lead in forming integrated partnerships with others involved and brought all parties together to form a wildlife and recreation management district, real progress might be made. The costs of producing the animals could be shared by the businesses and users who benefit from their presence. Park fees generated by visitors who demand large herds to view could be shared with the agricultural producers who sustain damages. Hunters who desire abundant populations for harvest could share in their production by paying fees either to the management district or directly to the landowners who produce and sustain the animals. Landowners who benefit from hunting income on their properties could share profits with neighbors who suffer losses without the opportunity for hunting income.

No management decisions would be made without considering the needs of all partners in the district.

The formation of local districts may seem overwhelming, but the idea has been successfully implemented elsewhere. Hunting districts and state-run cooperative hunting associations are used to address wildlife management in European countries, where the growing human population places major demands on the land and wildlife base (Decker and Nagy 1989). Bubenik (1976) states that the *revier,* or district, system in Europe represents the most efficient form of game management.

The Landcare program of Australia is one of the newest and most exciting programs to involve people in solving local conservation problems. Andrew Campbell (1994), in the introduction to his book about Landcare, describes the essence of the program as communities working together to solve their own problems and create their own opportunities.

In the twilight of the twentieth century, the need for humans to rethink the way we use the land, water, air and biodiversity which supports us has become pretty obvious.

The imperatives are economic, social and environmental. They are characterized by huge scale and technical uncertainty and the need for action is often urgent. It is easy to be overwhelmed by the size and complexity of the questions and we do not have neat, off-the-shelf answers.

One of the most pressing issues facing society over the next generation is how we produce food and fibre. Farming systems will need to support twice as many people by the year 2025 as they do today, hopefully with a much more equitable distribution of food. Yet, all around the world, farmers are under financial and social stress which, together with inappropriate technologies, have accelerated depletion and degradation of natural resources—soil, water, air, and biodiversity. Farming communities all over the world are in decline.

We want to introduce an exciting Australian phenomenon which shows how local communities, particularly but not exclusively rural communities, can get together to tackle their own problems. It is called "Landcare," a unique national program which is a partnership of government, farmers, conservationists, and community groups, and which has quickly grown to involve about one quarter of the farming community in local voluntary conservation groups.

If there is a word which emerges after four years immersed in Landcare it is potential. Landcare, by involving, encouraging and providing resources to committed people closest to the land, has the potential to underpin the evolution of new land use systems and new relationships between people and land, which build upon human resources instead of discounting them or seeing them as part of the problem.

Solutions for the twenty-first century and beyond will not be easy. The United States can capitalize upon its well-developed and, for the most part, excellent management system yet find ways to enfranchise the private sector. We can see the 60 percent of our land that is under private, nonindustrial ownership become the new frontier for wildlife and recreational opportunities or we can fight over ownership, management rights, and authority at the expense of wildlife, recreation, and the land. Leopold (1930) tells us to "make game management a partnership enterprise to which the landowner, the sportsman, and the public each contributes appropriate services, and from which each derives appropriate rewards."

CHAPTER 11

Exploring Our Wilderness Values through Recreation on Private Lands

A wide range of recreational experiences can be offered on private rural lands for the benefit of landowners and society at large. Visitors to rural areas find open space, pastoral settings, slower-paced lifestyles, contact with historical roots, and the peace and solitude that are conducive to art and music. Hunters and anglers learned long ago about the value of private lands for the kinds of recreation they enjoy. Today avid bird watchers, history and archaeology buffs, campers, hikers, rock climbers, and others are beginning to ask private landowners for access, too.

In general, landowners are slow to recognize the value of their recreational resources and to market them, perhaps because their rural lifestyles have traditionally led to the view that living on and working the land is hard enough work without interference from outsiders. They may think of their lifestyles as personally rewarding but not marketable. Leisure time is a true luxury for farmers and ranchers, and they may view recreation as curious, if not frivolous.

Yet there is ample evidence that rural recreation is indeed in demand and quite marketable. There has been a dramatic increase in the number of rural bed-and-breakfasts and in the number of farms and ranches that offer "working vacation" packages that allow their customers to participate in agricultural activities. Movies such as *City Slickers* show how people can live out their fantasies of the Old West. *Country* magazine and others reinforce the desirability of the rural life and give us a glimpse of a lifestyle that is rapidly being lost to urbanization.

Just what is the lure we find so attractive in our open spaces? Most people would describe the experiences they seek as being connected with the idea of wildness or wilderness, yet they may not think wilderness can be found on private lands. The term *wilderness* has come to represent remoteness, naturalness, and the absence of human impact. There are many private lands that fit this description but many more that do not. And yet even altered landscapes may fulfill the desires of many recreationists for what they would call wilderness experiences.

What Are Wilderness Values?

Wilderness means different things to different people (Clark, Hendee, and Campbell 1971). For some, it means vast reaches of land unchanged by human intervention. For others, it simply means any rural area or open space. Whatever our definition, we almost all agree that wilderness (or at least natural space) is valuable and needs to be preserved and appreciated.

This was not the case in the early days of North American settlement (Lucas 1964). The idea then was that wilderness was something to be conquered, not protected. The New England wilderness was thought of as hideous and desolate. Mountains and wildlands were detested. The landscapes considered attractive were the soft, fertile fields improved by human husbandry. Formal gardens laid out with architectural precision were preferred over naturalness. French voyagers in the eighteenth century called part of the present-day Minnesota-Ontario border area "le beau pays"—the beautiful country—but the area they described was not the lakes and rocky shores. Rather, it was the glacial lake plain farther west, with level, often open areas reminiscent of farmland.

Over time perceptions of wilderness changed. The idea that wilderness was a valuable resource in its own right, rather than just land to be developed, was probably tied to the closing of the frontier. As the country became urbanized, people felt cut off from the land and lost a sense of continuity and security.

Led by such figures as George Catlin, Henry David Thoreau, and George Perkins March, the public began to appeal for wilderness areas to be protected. The creation of Yosemite and Yellowstone Parks resulted from such actions. The legislative act that established Yellowstone stated that it would "provide for the preservation from injury or spoilation, of all timber, mineral deposits, natural curiosities, or wonders . . . and their retention in their natural condition." Notice that there was no mention of protecting the area from human access. As yet, the idea of wilderness did not mean large, roadless tracts as it does today. Likewise, the Adirondack Forest Preserve in New York was established as "forever wild" in 1885 mainly to prevent timber exploitation rather than to promote general wilderness values.

The setting aside of specific wilderness reservations began in the 1920s under the leadership of Aldo Leopold, Arthur Carhart, and Robert Marshall. Another surge came in the 1930s, followed by major designations in the 1960s with the passage of Public Law 88-577, the Wilderness Act of 1964. This formal designation stated that a wilderness is:

> an area where the earth and its community of life are untrammeled by man, where man himself is a visitor who does not remain . . . an area of undeveloped Federal land retaining primeval character and influence, without permanent improvements or human habitation, which is protected and managed so as to preserve its natural conditions and which:
>
> 1. generally appears to have been affected primarily by the forces of nature, with the imprint of man's work substantially unnoticeable;
> 2. has outstanding opportunities for solitude or a primitive and unconfined type of recreation;
> 3. has at least five thousand acres of land or is of sufficient size as to make practicable its preservation and use in an unimpaired condition; and
> 4. may also contain ecological, geological, or other features of scientific, educational, scenic, or historical value.

A number of researchers have questioned visitors to wilderness areas to find out what values they place on their experiences there. Some of the adjectives or attributes they mention include: achievement, autonomy, escape from pressures, reflection on personal values, relationships with nature, sharing/recollection (Brown and Haas 1980); aesthetic enjoyment of natural settings, challenge and adventure, emotional or spiritual experience, escape from the familiar, escape from urban stresses (Rossman and Ulehla 1977); natural environment, pristine, solitude, uncivilized (Stankey 1973); and primitive, remote, rugged, uncommercialized, wild (Lucas 1964).

Although people share basic ideas about the value of wilderness, they do not all agree on where the "wilderness" is or how much of it still exists. Visitors to the Quetico-Superior area (in Minnesota and Ontario, Canada) were asked if they felt they were in the wilderness and where they thought the wilderness began (Lucas 1964). Canoeists thought the wilderness was a much smaller area than other recreationists who did not spend as much time there. Similarly, car campers thought there was less wilderness than travelers who were just passing through. Some respondents said they never reached the wilderness, while others said it began more than one hundred miles back down the road in central Minnesota. A substantial number of respondents mentioned "the end of the blacktop" as the beginning of wilderness. Clearly these visitors did not share common ideas about wilderness. Their perceptions varied according to the types of recreation they were pursuing.

Catton (1969) found two motivational extremes between ardent mountain climbers and sedentary campers. The contrast between the mountaineers' quest for uncertainty and challenge and the campers' quest for privacy, freedom to socialize, and freedom from tension led Catton to suggest that visitors may have either specific or nonspecific purposes for their recreational experiences. They may want to pursue certain activities or they may simply want to be in a particular place. Other studies have shown that the satisfaction people have with their recreational experiences, when wilderness values are important to them, can be ranked along a continuum increasing from city parks through state parks to wilderness areas (Iso-Ahola 1980). When people find the wilderness experiences they seek, whatever their definitions of wilderness may be, they find those experiences valuable and rewarding. Needless to say, wilderness experiences are not found only in designated wilderness areas.

It is clear that private lands hold vast potential for providing recreational opportunities, whether people seek specific or nonspecific experiences and whether or not they desire the remoteness of true wilderness. Landowners can easily tailor the kind of recreation they offer to their own desires and resources and to the particular market they want to reach.

Still, we must address the factors that influence landowners' attitudes toward recreational development because these may limit the willingness of landowners to consider new enterprises. Shilling and Bury (1973) found that these attitudes fall into three broad areas.

First, the landowner's philosophy of land management is critical. Some landowners might fear that opening the land to public use would degrade its value for other purposes. However, recreation does not have to displace the agricultural activities on the farm or ranch, and it can be a source of much needed supplemental income. For example, cattle ranches in the West require large tracts of land, and the return on the landowner's investment is poor. Those same tracts of land, with or without cattle, can be turned into recreational assets. Another reason landowners may be slower to provide opportunities than the public is to accept them is that managing people and recreation is quite different than managing traditional agricultural production. New skills and knowledge may be required. For example, the landowner may need assistance with financing, market research, and promotion.

The second finding about landowners' attitudes is that they are concerned that their natural resources might be degraded from increased human use. They fear that visitors might cause fires, vandalize property, and litter. We have discussed these kinds of concerns in previous chapters. The key to dealing with these fears is good communication between the landowner and the customer. The landowner must let the customer know what areas of the property are off-limits and what activities or behaviors are not allowed. Recreational access is a privilege that can be denied to those who are unwilling to agree to stated expectations.

The third concern landowners have is that free access to public lands may reduce the likelihood that people will pay for access to private lands. The fact is, public lands cannot accommodate the demand for recreation that exists now, and that demand will surely grow in the future. Hunters and anglers have set a precedent by showing that recreationists will pay for the activities they desire.

A more legitimate concern might be how to structure the fees charged for various kinds of recreation. Generally, persons who want to take advantage of services such as modern campgrounds, roads, sanitation facilities, interpretive programs, etc., are prepared to pay extra beyond the basic entrance fee. Persons in this category also expect to share the area with others and are prepared for greater congestion. They may even desire association with other campers. Because people are confined to organized areas and because more people can be accommodated, the land is used effectively and efficiently for the revenue received.

If the property is large enough, as is the case with many western lands, some demand for less organized, more wilderness-oriented activities may be accommodated at the same time. Because the landowner must provide more space for those who want to hike, camp, climb, hunt, fish, or photograph wildlife in remote areas with few others, these people can expect to pay more than those who want to stay in campgrounds.

If the property is not large enough to offer both high- and low-density areas at the same time, the same land might accommodate each at different times of the year. Fees should reflect the exclusivity of the use in addition to services provided. Consequently, persons who want to spend two weeks on the property alone or with a small tent-camping party would be expected to pay more per person than those in high-density camping areas.

The management and marketing of recreation on private lands may be the challenge of the future. It may seem a new concept, but dude ranches in the West and farms offering vacations in the East have operated successfully for many years. The good news for landowners who want to start recreational enterprises is that there seem to be markets for all kinds of activities and experiences. There is no one model that must be followed in order to be successful. As our society moves further away from a close association with the land, people will value and pay for opportunities to experience their own ideas of wilderness.

CHAPTER 12

Toward a Personal Conservation Ethic

The outstanding scientiWc discovery of the twentieth century is not television, or radio, but rather the complexity of the land organism. Only those who know the most about it can appreciate how little is known about it. The last word in ignorance is the man who says of an animal or plant: "What good is it?" If the land mechanism as a whole is good, then every part is good, whether we understand it or not. If the biota, in the course of aeons, has built something we like but do not understand, then who but a fool would discard seemingly useless parts? To keep every cog and wheel is the Wrst precaution of intelligent thinking.

—Aldo Leopold, *A Sand County Almanac*

These words penned at the midpoint of the twentieth century can prepare us for the twenty-first century because we can never be too holistic in our approach to managing land, wildlife, and people (Benson 1993). Perhaps these thoughts will inspire your own conservation ethic.

Earth has its own life as an organism. It is a dynamic and diverse system that takes inputs from the sun's energy and the atmosphere and converts the heat and gases internally into interdependent plants, animals, people, and their by-products. The September 1989 issue of *Scientific American* reviewed environmental threats and strategies for dealing with them. Some of those threats have almost become clichés: human overpopulation, acid rain, ozone depletion, the greenhouse effect, desertification, poverty, political unrest, decay in the cities, misuse of agricultural chemicals, reduction of rain forests, the disappearance of wild places, and loss of biological diversity. Can twice the number of people depend upon the same environment in the twenty-first century? Will they expect higher standards of living, social and personal security, better education, more buying power, increased mobility, a greater appreciation for and access to natural environments, and other symbols of a developed life? If the expectations are met, what price will the environment pay?

Fortunately, the future may not be disastrous according to Greg Easterbrook in his book *A Moment on the Earth: The Coming Age of Environmental Optimism* (1995). He makes the assertions, with examples to justify his claims, that:

- In the Western world pollution will end within our lifetimes, with society almost painlessly adapting a zero-emissions philosophy.
- Several categories of pollution have already ended.
- The environments of Western countries have been growing cleaner during the very period that the public has come to believe they are growing more polluted.
- First World industrial countries, considered the scourge of the global environment, are by most measures much cleaner than developing nations.
- Most feared environmental catastrophes, such as runaway global warming, are almost certain to be avoided.
- Far from becoming a new source of global discord, environmentalism, which binds nations to a common concern, will be the best thing that's ever happened to international relations.
- Nearly all technical trends are toward new devices and modes of production that are more efficient, use fewer resources, produce less waste, and cause less ecological disruption than technology of the past.
- There exists no fundamental conflict between the artificial and the natural.
- Artificial forces which today harm nature can be converted into allies of nature in an incredibly short time by natural standards.
- Most important, humankind, even a growing human population of many billions, can take a constructive place in the natural order.

Easterbrook goes on to say:

Ecorealism will be the next wave of environmental thinking. The core principles of ecorealism are these: that logic, not sentiment, is the best tool for safeguarding nature; that accurate understanding of the actual state of the environment will serve the Earth better than expressions of panic; that in order to form a constructive alliance with nature, men and women must learn to think like nature.

First, it [ecorealism] is not a philosophy of don't worry, be happy. Second, ecorealism is not an endorsement of the technological lifestyle. Third, and last in the inventory of what ecorealism is not, ecorealism has nothing to do with a minor fad called wise use. The phrase "wise use" once had a progressive meaning in environmental letters, but in recent years has been expropriated by reactionary fundraisers. Today lovers of nature ought to have no use for wise use. . . . There was a time when to cry alarm regarding environmental affairs was the daring position. Now that's the safe position: People get upset when you say things may turn out fine.

Of course, society should *work* to make things turn out fine. Substituting the sustained use of resources for earth-threatening activities was not as important for past generations as it is now because there were always new lands to discover where new resources could be used and new social systems could be established. Those lands have now been conquered and those resources sometimes overused. Human populations continue to grow and make ever greater demands on the environment. Land managers, scientists, politicians, administrators, indeed all of us, must become better stewards of the environment.

Curing environmental and social ills would be easier if these systems operated in simple ways. They do not! Fortunately, the complexity and versatility of human and natural systems also contribute to their stability. Solutions to our problems will be found by accepting complexity and by working with the varying philosophies and paradigms that can often create roadblocks which retard human compassion, logical thought, and compromise. Finding the middle way through conflicting points of view will help society achieve good resource management through cooperative action. Discussion of some of these dichotomies follows.

The Intrinsic vs. the Instrumental Value of Environments

The debate over the intrinsic vs. instrumental values of nature is an example of a wasted argument that affects how and when resources are managed and used. It can only be true that our universe has intrinsic value, that environmental systems are valuable unto themselves apart from any human association or construct. To argue differently is merely an exercise. It is also true that humans attribute values to environments as a result of their interactions with them and the economic benefits derived from them. Those are instrumental values. Values are human constructs. Classifying and debating values is important, but there is the risk that dogmatic rhetoric mainly serves academics and may lead society as a whole into narrow thinking that exacerbates environmental dilemmas.

Most people do not know the meaning of intrinsic or instrumental values, but they know what life and land mean to them. They are influenced by their experiences, education, culture, and religion or lack of religion. People behave in ecologically acceptable or unacceptable ways and use these parameters to justify their behavior.

Whatever one's value system, it is important to take a pragmatic look at Earth and society, accepting that ecologically sound behavior is the only meaningful solution to sustaining life. Leopold (1949) writes that we should "examine each question in terms of what is ethically and aesthetically right, as well as what is economically expedient."

Single Species vs. Biological Systems Management

A debate has developed within professional societies about the proper ideologies for guiding the management of the environment and all its parts. Some advocate the management of single species, while others insist on multiple species management. The debate may lead to professional division (Wagner 1989) and duplication (Hunter 1989) rather than to the generation of healthy new ideas (Edwards 1989) or a search for the collective good ("Future of Wildlife Resources" 1989). Rather than holding on to narrow views, we must actively strive for environmental integrity. Whole systems must be protected, evaluated, and managed; yet, to manage and monitor whole systems we must understand and take care of the individual parts. There is need for both. Fragmented ecosystems and endangered species must receive intensive, short-term assistance more than ever before. At the same time, work must go on to protect ecosystems large enough to merit an extensive approach.

Large, nonfragmented landscapes are the objective, but where are they to come from? Most of our nature reserves, wilderness areas, and parks are ecological islands amongst a sea of human development, and they are too small to sustain their species indefinitely. Leopold's (1921) recommendation for the size of protected wilderness areas was based upon his sense of both ecological integrity and human values. He called for a continuous stretch of country preserved in its natural state that is big enough to accommodate a two-week pack trip. His recommendation went unheeded. Barely one-third of all our wilderness areas in the United States exceeds one hundred thousand acres, and one in six wilderness areas is less than five thousand acres—an area that can be crossed in a few hours

(Cole 1990). Since the opportunity to set aside new wilderness areas is rapidly diminishing, our best option for creating those large, unfragmented landscapes is to cooperate with private land managers adjacent to public lands.

Integrated partnerships among governments, private landowners, and communities become vital (Benson 1990, 1991a). One model is the Man and the Biosphere Program recommended by the United Nations Educational, Scientific and Cultural Organization (Peine 1984; Dyer and Holland 1988). This model proposes that there be undisturbed core areas of land surrounded by buffer zones where resources are used and managed by the people who live there, ringed again by more densely populated areas. The model has logical merit, but to implement such a system would require the cooperation of the people. Human cultures everywhere will place their needs ahead of environmental needs unless they understand the consequences of their actions. However, if they are informed, included in decision making, and empowered to take part in resource management, they can become good stewards of the environment.

Unequal Distribution of People and Resources

Decisions to use or preserve resources are based, in part, upon human needs. It is unlikely that those whose needs are met by the human and natural resources at their disposal will sacrifice greatly in order to accommodate the needs of the poor. It is equally unlikely that the poor will achieve equal access to human and natural resource capital. Resources are consumed most by the wealthy, yet the numbers of poor and their demands for resources grow as well. Some lands are rich in space and other resources, yet have relatively few people. Other lands are filled with people beyond the environmental capacity of their borders. This is not likely to change. Resources have never been available equally, and if they were, it is doubtful that the environment could survive. What would the world be like if all persons had automobiles, electric conveniences, medical care, and attorneys? Human and natural resources would be depleted.

The largely developed Northern Hemisphere has a tremendous impact on the resources of the people of the largely undeveloped Southern Hemisphere. The North may owe an ecological debt to the South that is several orders of magnitude greater than the foreign debt currently imposed upon the South (Altiere 1990). The North is obligated to show care for the people and resources it affects, wherever they are. After all, Earth is one functioning organism.

All people should care for the environment. Natives of the Amazon should live within the sustainable means of their resources and practice the conservation methods appropriate to their society. So, too, should societies in developed countries live within their environmental means and those of other areas they affect. When individuals and societies learn to live within our world's environmental means we will make wise decisions about managing our natural resources.

Global Perspective vs. Local Perspective

A holistic worldview and a holistic local view are equally important. The world organism is merely a combination of its component parts. Each affects the other, so they cannot be separated. Each human contribution to improving the Earth organism is significant.

Resource Use vs. Resource Preservation

Land and wildlife management have evolved in stages from overexploitation to protection and preservation without our fully understanding and allowing for the inherent needs of species and ecosystems, communities and human societies. When the needs of the environment and the needs of people conflict, the response has too often been to sacrifice one or the other. Management conservation, as an alternative to either extreme, is often overlooked.

As a result, our approach has been to set aside fragments of environments as protected areas in parks, nature reserves, forests, rangelands, and waters. In retrospect, too few areas were protected in this way. Understanding and wise decisions came too late to save much of the habitat of species such as the bison in North America or elephants in Africa, whose populations could number in the millions if there were places for them to inhabit. Governments around the world could not set land aside fast enough to protect plant and animal systems from the pressures of human population expansion.

Massive sociological and ideological shifts will be required to change the trend toward carving up the world's remaining resources and distributing them among the fortunate. Wildlands that still exist should be protected wherever possible, but building political and actual fences around them and then prohibiting human activity is an impossible and socially divisive approach. Some lands and waters may need to be protected from all human intervention, but most environments will need intensive management. We cannot rely on governments to do all that needs to be done. Governments and the private and communal sectors must work together. That is the way to protect our natural resources and use and manage them wisely in modern times.

Resource protection, resource conservation (wise use), and resource steward-ship (management) are all appropriate phrases to describe how humans must interact with their environment. Our demands for food, fiber, energy, water, and other resources need not destroy the world we live in. Renewable natural resources can be used wisely and sustained if all the consequences of using them are considered in the decision-making process. Nonrenewable resources must be managed with extreme caution. We should use them only if ecosystems can func-tion without them or their by-products. We must be careful in the future to avoid the mistakes of the past. In some situations naturalness and wildness will have to be sacrificed, but in others careful management can preserve natural ar-eas even while making use of the resources they contain.

Public Ownership vs. Private Ownership

We have spent considerable time discussing the conflicts between public and private ownership of land and resources. Another way to place this dilemma in context is to connect it to the global vs. local dichotomy. Governments rightly serve global interests, while individuals have a natural and immediate interest in their own local situations. The role of both is important, but neither can man-age our resources alone. Not only should the public sector be entrusted to care for the holistic needs of society and the sustainability of the environment, but it must also trust, enable, train, and support individuals to join in conservation and resource management.

The guiding principle for all must be stewardship of resources. Economic rewards and society's gratitude should go to persons who practice good steward-ship, while punishment and disgrace should be the fate of those who abuse our resources.

An Agenda for Conservation and Resource Management

There are at least two ways of approaching resource management, aside from the option of doing nothing. One way is to let nature take its course, with human intervention taking place only when needs arise. The second way is to use our human intelligence to manage the Earth wisely, anticipating problems and find-ing solutions. Surely the second is the better way. The goal for resource manage-ment should be an infinitely sustained environment. We humans have the power to work toward that goal. Not only is it the right thing to do, but it is in our own selfish utilitarian and aesthetic interests. In our universities, governments, and research organizations we have the institutions that can help us. Tolba (1990) foresees that their function will be to gather knowledge about air, land, water, plants, animals, and humans and then communicate that knowledge to the rest of us so we can make wise decisions and take proper actions. Careful monitoring of barometers such as biological diversity, minimum viable populations, habitat fragmentation, keystone species, extinction rates, population declines, and encroachment of invader plants and animals that affect indigenous species will make it possible to warn of danger in time to avert disaster. However, taking action when the warning is sounded is not just the government's job—it is people's job. According to Vrana (1990), the role of federal, state, and local conservation agencies will change. The federal government will develop databases of information, environmental standards, technical specifications, and guidelines. Environmental actions will be carried out by local agencies, with state oversight and a watchful citizenry.

Holistic Management of Natural Resources

In the future a new way of thinking must guide our management of natural resources. Separate disciplines must come together to devise holistic plans that can be carried out, with guidance and assistance, by individuals and communities. We have to devise solutions that translate to concrete actions so that every person can contribute to environmental stewardship.

Accepting Our Role as Environmental Stewards

Conservation really depends upon public sentiments. Do we humans have the will to make the changes that are necessary? Berry (1989) writes, "We must acquire the character and the skills to live much poorer than we do. We must waste less. We must do more for ourselves." All of us must work toward making fewer demands upon the Earth. Orr (1990) suggests that an unlikely enterprise awaits if humans expect to redesign the Earth to account for insatiable appetites, economics, technologies, and the propensity to breed. We need to train ourselves to live within our environmental means. Governments, educators, media, and the business community should make environmental literacy and personal conservation world priorities, teaching society to live in harmony with the environment.

As we approach the twenty-first century, we can trace the global development of an environmental ethic and have hope that more progress will be made in the years to come. The 1950s were a time to recover from World War II and concentrate on restoring normal life. The new industrial revolution and international baby boom continued. In the 1960s the "silent spring" (Carson 1962) of environmental degradation was broken by an infant environmental movement in North America. Young people protested and laws were changed because they had to be. Environmental progress began. The 1970s brought Earth Day and international endangered species conventions. Governments took time to work on their environmental agendas, sorting through conflicts and beginning to educate the people. Some of us read about conservation ecology (Cox 1969) and the American crusade for wildlife (Trefethen 1975), while most Americans gleaned knowledge about the environment from conservation organizations and the mass media.

Children of the 1950s became the leaders of the 1980s, and environmental issues were an accepted part of life. The new younger generation picked up litter and cried out for the rain forests and animal rehabilitation, while the establishment worried about holes in the ozone, acid rain, and global warming. Academics wrote about fragmented habitat and decreasing biodiversity, while land and wildlife agencies began to talk more holistically rather than focusing on individual plant and animal species. In the 1990s there appears to be a sincere and institutionalized movement within our society to work proactively with the environment rather than neglecting it or reacting to it. Yet words are still more common than actions.

Our hope for the decades to come is that individuals and societies will learn to live by an environmental ethic, understanding that life cannot be sustained as we know it and want it without making reasonable sacrifices and accepting responsibility. The conservation of natural resources and the management of wildlife species are not the jobs of governments alone; they are our jobs too. The oceans have already been crossed, the continents explored, the mountains climbed, and the lands altered. There is no other side of the mountain. We must preserve what we have—all of us—as partners in stewardship.

Wildlife Enterprise Analysis

Hunting, fishing, and other recreational enterprises can become an integral part of your farm and ranch management plans. Like other income-producing ventures, they require capital, time, planning, thought, and management. The amount of effort required of you will depend on the size of your operation, the type of management you use, and the program you choose. The work sheets that follow will help you decide whether to begin a recreational enterprise and lay the groundwork for future decisions.

The first step is to examine your options. Ask yourself:

What Do I Want to Manage?

__ Big game (deer, elk, pronghorn, etc.)
__ Small game (turkey, pheasant, quail, grouse, dove, etc.)
__ Waterfowl
__ Fishing (warm water, trout)
__ Farm and ranch vacations
__ Historical visits
__ Bed-and-breakfast
__ Guiding and outfitting
__ Bird-watching
__ Shooting preserve
__ Wilderness experiences
__ Rural experiences
__ Camping
__ Photographic safaris

How Will I Benefit?

__ Money

__ Return "in kind" (services, maintenance, patrolling, etc.)

__ Better control of people who use my land

__ Reduced vandalism

__ Social benefits (pride in full use of resources, respect of neighbors, etc.)

__ Advertising and public relations benefits

__ Ethical and philosophical benefits (doing what I think is right)

How Will I Organize?

__ Management by myself

__ Management by my employee(s)

__ Management by a professional wildlife recreation consultant

__ Lease to a large sportsmen's organization

__ Lease to a professional wildlife recreation consultant

__ Lease to a local club

__ Lease to an individual hunter or guide who will organize clientele

__ Lease to friends or relatives

__ Trade access for services, goodwill, other benefits

__ Cooperate with other landowners

These questions may help you think of other choices. Now choose one set of options, visualize how it would affect your present operation, and answer the following questions:

Is the Project Workable?

__ Do my land and water bases make the operation possible?

__ Is there demand?

__ Would the timing of work in the wildlife operation fit in with my other operations?

__ Is the hunting season long enough to be profitable?

__ What should I charge?

__ How many recreationists can I accommodate?

__ How productive is my land?

__ Are experts available?

___ What laws and regulations affect my proposed operations?

___ Where will I find my clients?

What Will the Costs Be?

___ What front-end development costs are there (structures, utilities, habitat improvements, facilities, roads, office space, attorney and accountant fees, insurance, shooting preserve license fees, etc.)?

___ Will cash flow require operating loans?

___ Will I have to borrow money for capital investment?

___ What are current interest rates?

___ Do I have an assured loan source?

___ Are there additional costs to other operations as a result of the wildlife program?

___ How much capital is required?

___ How much could I make with that capital if I invested it elsewhere? Potential investment income is a cost of the wildlife operation; subtract it from gross income.

___ What percentage of expected revenue should be assigned to risk and error?

___ What cost value should I ascribe to my labor and my family's labor?

___ Will there be employee costs? How much?

___ What are the total financial costs?

___ What nonmonetary costs are there (loss of solitude, dealing with people, additional responsibility, etc.)?

What Are the Likely Returns?

___ What are expected gross revenues?

___ Will there be savings in other operations as a result of the wildlife and recreation program (e.g., damage prevention, maintenance, and construction paid by recreation program)?

___ What advantages will recreationists provide me? What is the value?

Subtract total costs to get net revenue. Divide annual net revenue by equity to obtain annual rate of return on capital.

___ Are there tangible benefits?

___ What are the total benefits?

How Do I Compare with the Competition?

Make a comparison of what you plan to offer with that of similar enterprises in your area or in other parts of the state and country (see table A.1).

Table A.1. Comparing My Operation with a Competitor's Operation

Item for Consideration	My Operation	Competition
Distance to nearest clientele center		
Dependability of transportation: Air Surface		
Size of clientele group		
Size of operation		
Distance from competition		
Cost for client's trip @ $0.25 per mile and $75 per day		
Services offered: Guides Vehicles Dogs Horses Beverages Food Lodging Camping Harvest costs Trophy costs Tournaments and prizes		
Number of clients who can be accommodated at one time		
Number of clients who can be accommodated annually		

Quality of experiences

Aesthetics of property:
>   Animal populations
>   Size of animals
>   Hunter success rate
>   Time required for success
>   Scenery
>   RV and tent-camping facilities
>   Cabins
>   Wilderness opportunities
>   Surrounding tourist attractions
>   Entertainment
>   Daytime activities for spouses, families
>   Bird-watching quality
>   Other recreational opportunities

Prices charged

Should I Do It?

By now you have gathered a lot of facts and made some assumptions. To make some sense out of it, complete table A.2. In the "Response" column put +1 if the answer is yes, 0 if the answer is maybe, and -1 if the answer is no. Some questions will affect your operation more than others. Place a number from 0 to 5 in the "Importance" column, with 0 indicating no importance and 5 indicating great importance.

To summarize your answers, multiply numbers in the two columns and place the answer in the "Weighted Value" column. Add the weighted values. Zero is neutral. A negative number is a "no" answer, and a positive number is a "yes" answer. Possible scores range from -100 to +100.

A weak affirmative score probably would not be enough to justify beginning a recreational enterprise.

Table A.2. Decision Criteria for Beginning a Recreational Enterprise

	Response to question	Importance of question to decision	Weighted value
Do you enjoy working with people?			
Do you presently have problems with recreationists?			
Do you presently have property damage from recreationists?			
Do you anticipate a decrease in property damage as a result of a commercial recreational business?			
Do you enjoy the challenge of different and new businesses?			
Do you have employees who can insulate you from contact with recreationists?			
Do you support bird-watching or other nature study?			
Do you support hunting?			
Do you enjoy strangers using your land?			
Do you enjoy meeting people and making new friends?			

Is the operation
financially worthwhile?

Is financial return
your objective?

Have you identified a
dependable clientele?

Are there other benefits?

Are the nonmonetary costs
small?

Do nonmonetary costs
balance nonmonetary
benefits?

Are there positive
effects on your other
operations?

Do your family, friends,
employees, and neighbors
favor the idea?

Do you favor it?

Total

Making the Decision

Review your options. If you have changed any options or want to examine a different set, answer the questions again to make sure new options have not affected the score. If you have no more questions, you have arrived at the time of decision.

It is hard to apply a score accurately to intangible factors. Do not use that number alone for your final decision. Reconsider your family plans, tax status, estate planning, desired lifestyle, and other intangibles to make sure they had enough influence.

What Did You Decide?

__ Yes (which option?)

__ No

__ I will try it for a while.

Note: This material was modified in 1988 by Delwin E. Benson and published in *Wildlife and Recreation Workbook* XCM-117, Colorado State University Cooperative Extension Service, from work sheets prepared by Lee Fitzhugh, California Cooperative Extension Service.

Lease Agreements
and Hunting Systems

Ideas Basic to Most Leases

This is a listing of components that should be included in any lease. It is a useful
guide, but you should consult with your attorney about specifics pertinent to
your situation. Be sure to speak directly with customers about provisions of a
lease. Do not rely on them to read and understand all that you have included.
Maintain good communication throughout the term of the lease.

1. Identify the lessor and lessee. (You may want to incorporate your recreational
    enterprise separately.)
2. Include the date of the agreement and the location where it is signed.
3. An agreement to arbitrate differences may prevent many disagreements from
    occurring. Some leases specify that the landowner is the final authority in re-
    solving agreements so that the rights of the lessee do not interfere with the
    landowner's other operations.
4. Conditions for the cancellation, renegotiation, or renewal of the lease may
    specify the conditions under which these transactions may occur. For example,
    the lease might be canceled if either party desires, or if there is cause, e.g.,
    failure to perform according to the lease provisions, unsafe or unsportsmanlike
    conduct, property damage, leaving gates open, failure to obey state laws and
    county regulations, or other objectionable behavior. The lease should specify
    the manner in which the notice of cancellation or renegotiation is to be con-
    veyed and the disposition of the fee upon cancellation. An automatic termina-
    tion or renewal clause may be desirable.
5. Security deposits may be required in advance, either to hold reservations or for
    repair of property damage. Deposits normally are applied to the lease payment
    or are returned unless provisions of the deposit clause are violated. In order to

retain a deposit for damage repair, the landowner must be able to demonstrate that the lessee was responsible for the damage. Conditions for the return of the deposit should be specified.

6. It is important to include a description of the present condition of the property after the property has been inspected by both parties and its condition agreed to. This might be handled in a separate document, especially if structures or other improvements are involved. It helps to have pictures of the interior of cabins and other structures where damage may occur. Take photographs during the joint inspection. Include the lessee in some pictures and give the lessee a copy.

Description of the Lease

7. Specify exactly what privileges are being granted to the lessee (e.g., the right of access for the purpose of hunting or bird-watching) and what money or services (e.g., road and water tank maintenance, patrolling during hunting season) the landowner will receive in exchange. Also specify any restrictions on access, such as the requirement that the lessee be accompanied by a guide or that the lessee not enter certain areas of the property. If access is to be guided, include details such as the number of people per guide, number of days access is allowed, any other scheduling provisions, and fees. Services to be provided by the landowner or the lessee should be detailed with dates, locations, and manner of performance.

8. Acreages and boundaries should be clearly explained. The lease may not include all of your property. You may want to exclude your living area, certain pastures or farming areas, and areas leased to other individuals. Buildings and other structures, blinds, vehicles, campsites, water, sanitary facilities, and other significant features should be identified. A map showing these features is sometimes part of the lease.

9. If the lessee is to construct or maintain improvements, mention standards for such work.

10. The lease should state whether the landowner has the obligation and right to police against trespass and hunting without permission, or whether this obligation is being transferred partially or completely to the lessee. If the lessee is to patrol to prevent trespass and hunting without permission, the lease should state the areas to be patrolled as well as the dates and times of day the patrols are to take place.

11. If items or services are to be provided either by the lessee or lessor, the cost of

such items or services should be reserved to one party or another. For example, the landowner may buy the materials to be used in improvements with the construction performed by the lessee, or the lessee may be responsible for utility costs in a cabin.

Protection for the Landowner

12. Rights and premises retained by the landowner should be specified. For example, you may want to retain hunting rights for yourself and your family or for another lessee. Consider whether you want to retain access and other rights for yourself and your employees so that you can conduct other business operations. If livestock are to be grazed in the leased area, that right could become a point of contention unless reserved. If lessees are allowed to make improvements, consider whether you want to retain the right to choose improvement locations.
13. A provision for indemnification of property damage is sometimes included.
14. If authority is granted for the lessee to act for the lessor, whether in financial transactions, improvements to property, or patrolling, the limits of that authority should be set forth.
15. Besides the standard waiver of liability, you might want to incorporate principles of firearms safety or require successful completion of hunter- or water-safety classes, or perhaps prohibit driving, boating, or hunting while under the influence of alcohol or drugs as a condition of the lease. Liability for wildfires also should be covered.
16. Sometimes the lease includes an acknowledgment of risk. This can be handled by describing the nature of the property (e.g., wildland, barbed-wire fences, logs, poison ivy, wild animals, livestock, uneven terrain, and other hazards) and the nature of the activity (e.g., hunting, hiking, birding, boating, and other recreation), which the lessee would acknowledge in signing the lease.

Restrictions on Lessee

17. Identify whether or not subleasing is allowed and whether or not there is a restriction on the daily or total number of guests, family members, and recreationists allowed. Consider whether lessees should be allowed to assign or transfer their interests in the agreement without the owner's consent, except as specified.
18. Lessees and guests should obey all state and federal laws and regulations, in-

cluding game laws and laws related to fires. Special speed limits and other traffic regulations are established in some leases.

19. Plinking, target shooting, and sighting-in of rifles may be allowed under certain conditions or may be prohibited.

20. The lease may state how the lessee is to care for the property. Examples might be procedures for opening and closing gates or caring for water sources and tanks, or a clause stating that the lessee will not litter, etc. The lease should stipulate if and where the lessee is allowed to build open fires or use charcoal, gas, or propane stoves. Some landowners levy fines for violations of these provisions.

21. The landowner may require that the lessee give notice before each trip to use the property.

22. If the lease is with a group, some landowners contract with one individual who is responsible for the actions of the group. If so, the lease might state that the lessor reserves to the lessee all responsibility for administration and enforcement of the lease with the group represented.

Hunter Management Provisions

23. The species, sex, and number of animals that can be hunted should be specified, within the limits of state and federal laws. This information also should be part of the annual renewal of a long-term lease.

24. The lease should require lessees to record, report, and/or show you their game for an accounting of species, sex, age, and numbers. The landowner may specify firearms to be used, methods allowed (dogs, blinds, vehicles, calls, and other techniques), and special tags required.

You will undoubtedly need other provisions for your particular situation, while some of those listed may not apply. In leases there is an optimum balance between flexibility and specificity. You and your attorney should try to find that balance based on what is best for you.

Types of Fee-Hunting Systems

There are four basic types of fee-hunting systems: permitting, leasing, commercial membership enterprises, and direct ownership. They sometimes overlap, but each has merits to consider.

Permit Hunting Systems

Permit hunting systems were instituted primarily by timber companies in the South. These companies had to consider the public relations aspects of charging for hunting and allowing access. There were early concerns about local reactions and the fear of disgruntled residents setting fire to the timberlands.

The daily permit program designated certain days for hunting, and the manager set up a certain number of hunts on those days. Permits could be bought for individual hunts or for all the hunts in a season. Another variation, called the party permit system, applied primarily to deer hunting. A tract of land was reserved for a particular party to hunt on a certain day. The hunting party was allowed to hunt the area by themselves. Fees were assessed in two ways: 1) each member was charged a certain fee, and a minimum number of hunters was required; or 2) each party was charged a flat fee to hunt the area regardless of the size of the group.

These systems evolved during the early exploratory efforts to manage permit hunting. They were never widespread and were principally found in states with long deer seasons and liberal bag limits, e.g., Alabama and South Carolina. These systems simply were too expensive and required too much labor for the revenue they generated, and they soon lost favor with both landowners and sportsmen.

Seasonal permit systems were also tried. Landowners charged individuals fees for hunting the entire season. Some landowners (mainly timber companies) charged variable fees. For example, a county resident could hunt small game on land a timber company owned in his county for one amount and could hunt all game for a larger amount. Still larger fees were charged to hunt on all lands the company owned in the state. Sometimes young people under sixteen were charged lower fees. Some companies still operate systems like this but have simplified their pricing to one or two levels with permits made available by mail or through designated agents (stores, district foresters, or offices, for example). Some seasonal permit systems have evolved into cooperative agreements with state wildlife agencies. Often the state will provide technical advice and assist with law enforcement. The state will either lease the land from the company for a fee or be given the lease for free. The timber company may receive revenues from the state lease or by selling season permits directly to individuals in order to provide public hunting opportunities in cooperation with state wildlife agencies.

One method used to initiate a permit program on land with little game is to post such property against trespass and implement conservation practices. This reduces game poaching, increases wildlife populations, and makes it apparent to the public that the company has gone to obvious expense to improve the hunting. Thus, when a fee program is introduced later, the landowner can better justify charging for hunting privileges.

ADVANTAGES AND DISADVANTAGES OF PERMIT SYSTEMS

Permit programs have several advantages. They normally provide recreation to a larger and more varied group of people; they allow landowners to increase or decrease the number of permits issued to increase or decrease the harvest of wildlife; and they can provide revenue to offset the costs of wildlife and hunters. There are also several disadvantages. They require higher administration, labor, and other operational costs; there is uncertainty as to the amount of revenue that will be received; and there may be less control over the quality of hunters on the property than with leasing, thereby increasing the possibility of conflict with other permit holders and/or neighboring landowners. While hunting leases in the South have increased, use of the permit hunting system has decreased.

Hunting Lease System

Leasing is the most popular fee-hunting system. However, more than half of all leasing in the United States occurs in Texas, Georgia, Louisiana, Virginia, Alabama, and Mississippi. Most southern leases are on a seasonal or annual basis, with property owners leasing to clubs, individuals, and companies. Most southern timber companies prefer to lease to local residents as opposed to distant urban residents in an effort to create better local public relations and reduce public reaction to income-producing programs. Day leasing is often found close to population centers and attracts a greater variety of clients. This system offers affordable opportunity to the casual hunter and capitalizes on a larger sportsmen market. With day leases advertising expenses are usually a larger part of total expenses than with seasonal or annual leases. It is also more difficult to predict income until the business is firmly established.

Many hunting clubs invest large sums of money in road development, culverts, fences, patrolling, and full-time caretakers that benefit landowners. Long-term leases tend to encourage this kind of investment because the lessee has time to recover his investment. With long-term leases landowners are more apt to offer a variety of services such as blinds, food plots, game processing, etc. Charges vary accordingly.

The majority of hunting leases in the United States appear to be annual. Lewis (1965) found that most hunting leases in Louisiana are annual, and McCurdy and Echelberger (1968) found that most goose and duck hunting leases in Illinois are annual. In the early stages of a leasing program annual leases can be justified, but after a cooperative landowner-sportsman relationship has been established longer leases probably will provide both parties with additional benefits. Some landowners are setting up leases for five years but are receiving payment at the beginning when the lease is signed. Sometimes the lease fee is discounted slightly for a longer-term lease.

LEASING STRATEGIES

Although many leases are with local residents, many are not. Because so many urbanites are seeking improved hunting opportunities, landowners are now advertising in virtually all major city newspapers across the Sunbelt and in such national publications as the *Wall Street Journal.* While some leaseholders are residents of the state where the property is located, others are from adjoining and distant states.

Some landowners are using brokers to find lessees. The broker agrees, for a negotiated fee, to find buyers at a price that is acceptable to the landowner. The broker advertises and promotes the hunting lease until an acceptable party is found. Brokers may also deal in real estate or may be hunting consultants who operate much like travel agents. They may even provide guides and outfitters for their clients. Brokers charge from 6 to 15 percent more if their responsibilities include managing or administering the lease. In some cases brokers will purchase leases and then sublease the hunting privileges to individuals or groups. Landowners who are not familiar with operating a recreational enterprise might find advantages in working through a broker.

ADVANTAGES AND DISADVANTAGES OF LEASING SYSTEMS

Advantages of leasing programs include better control over the land, the leaseholder's concern with caring for the land, better cooperation by hunters with work crews, larger game populations, and better-quality recreation for the optimum number of people the land can support. Other advantages include a minimum of management expenses to the lessor, being able to negotiate with only one person or board of directors over the leasing of the property, and advance knowledge of the income that can be expected. Disadvantages of leasing include: the possibility that some members of hunting clubs will begin to feel they own the land and may at times disagree with or interfere with an owner's land-management policies; the desire of some clubs to build clubhouses and put in utility connections (which may cause landowners to fear eventual easement problems from these activities); and the possibility that a hunting club will not have enough members to harvest the number of animals required for population improvement.

SHORT-DURATION HUNTS

There are various kinds of short hunts offered on some properties. With one-day hunts the landowner provides few, if any, services or amenities. The fee is fairly low and covers basic access. The type of hunting may be restricted.

Three-day hunts allow the hunter to use the property for a longer period and often include more amenities but minimum services. The rights to hunt are usually restricted as to the type and amount of game that can be taken.

With a package hunt, the landowner provides the maximum of services and amenities for the hunter(s). This may include high-quality housing, good food, and access to higher-quality game. A guide may be provided, or at least the hunter is given some idea of the best times and places in which to hunt. Package hunts usually are more expensive but also give the hunter a more satisfactory experience.

Commercial Membership Enterprises

This type of fee-hunting system purveys high-quality recreation to the person willing to pay a relatively high membership fee. The landowner sells or leases the right to hunt and/or fish through an annual membership fee and normally spends a considerable amount of time and money in managing the enterprise for the production of wildlife. Membership fees generally range from one thousand to five thousand dollars per year. Deer, ducks, turkeys, and quail appear to be the most popular game animals associated with commercial membership enterprises. The owners of commercial membership enterprises normally blend farming, ranching, and wildlife conservation practices in a complementary way. The advantages of this system to landowners include higher income than with other systems, some advance knowledge of potential annual income, and its complementarity with other agricultural enterprises. Disadvantages include the large capital investment necessary, the high administrative and operating costs, the strict location and managerial requirements, and the smaller (but perhaps potentially more profitable) sportsman market.

Land Ownership by Membership

A relatively new development in the South does not involve fee income but the outright purchase of timberland by wealthy individuals for hunting purposes. One recently advertised tract listed membership cost as $105,000 each, with financing available. A number of these developments have occurred in recent years, with membership prices ranging from $35,000 to $115,000. Usually the legal entities that hold these properties are limited partnerships or Subchapter S corporations. They are being marketed by the real estate industry.

ADVANTAGES AND DISADVANTAGES OF OWNERSHIP BY MEMBERSHIP

This system vests full simple property rights in sportsmen, which allows them to alter the landscape specifically for wildlife and conservation and to gain the tax advantages and federal and state cost/share program assistance that are available to other landowners. It ensures that the land will be dedicated to conservation virtually in perpetuity, either through conservation easements or because it is unlikely that the membership will desire to change its use of the land. However, the sportsman must make a large financial commitment to purchase the land and pay continuous membership fees to operate and develop the land for wildlife.

LAND OWNERSHIP BY INDIVIDUALS

The market for wildlife properties is increasing. One reason, of course, is that the number of affluent Americans has grown. However, even those with moderate incomes are becoming concerned about the environment and want to escape urban life to enjoy a rural lifestyle. Purchasing land for its aesthetic value, and its wildlife value in particular, may be a growing trend in the future.

Forming Wildlife
Management Associations

Managing for wildlife on small acreages is a special challenge. Wildlife is not bound by barbed wire fences or property boundaries on plat maps. Proper management for healthy wildlife populations and habitat depends largely on the actions, interactions, and attitudes of one's neighbors. Even larger tracts can benefit when neighbors cooperate. Imagine a line inside your property one-half to three-quarters of a mile from your boundary fence, and that will give you a good idea of the area on which you share wildlife with your neighbor. Cooperative wildlife management is already successful in some areas, and it is a concept that will move us toward a more holistic approach in the future.

The formation of a wildlife management association begins with talking to your neighbors about the idea. They may react negatively at first, but persist and remain positive. Give neighbors a chance to talk to others, and you may find that interest in forming a co-op will grow. Because most people cherish wildlife and want it to thrive, most landowners eventually will be interested, especially if others are willing to join in. In fact, a common problem is establishing boundaries to the co-op in order to separate one group from another. Don't worry about this initially. Many wildlife associations consist of numerous small groups that sprang from one source.

One way to get started is to invite your neighbors to an informational meeting. Invite a biologist, the local game warden, or your county Extension agent to speak to the group. These persons can explain the benefits of forming an association and will be glad to do so because such associations can also assist them.

Also invite absentee landowners and their land managers, and give them ample notice so that they can arrange their schedules. Absentee landowners will be very interested in wildlife management on their lands and often are key participants in associations. Allow a couple of hours for the meeting. Discuss the wildlife management program and forming an association, and have a time for questions and answers.

There are as many types of cooperative landowner agreements as there are wildlife management associations. Your guest speaker can give examples of several types. Typically, the agreements are not legally binding (so long as no money is involved). However, landowners enter into the agreement with an expressed intent to comply in good faith to the guidelines the group adopts, and to get recreationists on their lands to comply. These guidelines are derived from a management plan developed by the group in cooperation with a wildlife biologist.

The program should have an effective date (start date) and run three to five years. All parties should agree to participate in collecting data and observations. Long-term holders of recreational leases are usually interested in assisting.

Many organizations choose to incorporate enhanced law enforcement. This could be similar to a neighborhood crime watch program. Landowners may look after one another's property, post association lands, and even require recreationists to produce permission cards (this has been especially effective near urban areas). Your local game warden can provide valuable assistance by explaining laws and the type of information that is needed when violations are reported to local officials or to state wildlife agencies. In areas where trespassing, poaching, and night hunting are common, associations have succeeded in reducing such illegal activities. The obvious benefit is that the landowners regain control of their lands and can manage for wildlife. Legitimate hunters and holders of hunting leases not only approve of these activities but often are the most active participants.

Enforcement options should be discussed with your local game warden, and wording in the association agreement should be coordinated with your county or district attorney and sheriff's department. Remember that the agreement is only a guideline, and it can be simple or complex depending on the planned activities of the group. Something as simple as the following statement will work: "We the undersigned agree to the stated purpose of the group, and will attempt to comply in good faith with the management recommendations the group adopts, and will attempt to have our participants do likewise, and will otherwise endeavor to comply with sound game and wildlife management techniques." The purpose can then be simply stated. It should relate directly to your goals.

State fish and wildlife agencies will support your efforts to properly manage the natural resources of the state. These departments may supply printed certificates of membership to each member of the wildlife management association and may recognize outstanding wildlife and habitat work by private landowners. Agencies almost always provide assistance to persons wanting to include wildlife management in their present or future land practices. These advisory services are provided free of cost to cooperating landowners and land managers.

The joining together of a group of landowners for a common cause such as wildlife management is a major endeavor. The development of a program that involves various owners, operators, recreationists, and land unit sizes under different management philosophies can be challenging. The most important thing is that co-op members continue working toward the goals and objectives they have agreed upon. Neighbors who live close together and see each other on a regular basis can usually work out details concerning operational activities. Absentee landowners may have more difficulty relating to other members because they see them infrequently and may have different perspectives about ranch or farm operations. For example, one landowner may see the co-op as a means of increasing income, while another may be more concerned that the endeavor increase recreational activities. To avoid pitfalls caused by different perspectives, it is important that the members mutually agree to the goals they would like to accomplish.

Another problem may arise when some landowners within the management area are not members of the association. Normally, one of the goals of the association will pertain to harvest and management guidelines aimed at increasing the wildlife species to be managed. If these guidelines are not followed by a few nonmembers whose lands are within the boundaries of the co-op unit, the effectiveness of management on association lands may be reduced. However, the benefits gained from sound management practices should encourage nonmember landowners to join.

Those co-ops formed by landowners (and supported by users) who are truly interested and share a common goal have an excellent chance to succeed. The needs of the membership are best determined by the members themselves. Some of the largest acreages now being operated under one basic management plan were initiated by one or two ranchers. The adjoining neighbors and recreationists were able to see the benefits from such management programs and then started programs of their own.

Individual landowner and recreationist involvement is very important in allowing all members the opportunity of contributing to the success of the organization. Rotating officers of the co-op will allow responsibilities to be shared by different members who may introduce new ideas. This hands-on mode of operation may prevent members from feeling that they are left out of the decision-making process. Sharing the workload will give all members a chance to participate in co-op operations and make them feel truly important to the association's success.

Although management recommendations may be made for the entire area, landowners remain in control of their own operations. Good relationships with recreationists are important to maintaining a strong and active association. In the first year of the association, landowners and users will agree on the terms of the lease, the harvest quotas, and management plans for that year or for several years. The agreement should not be altered on short notice. Good communication is important so that recreationists can participate fully as members of the association. Users should be aware of the organization's goals and the goals of the ranches they are using.

Members should set attainable objectives and not expect immediate results. Normally, at least three years will be required before management results are evident, and a five-year period is required for maximum production. Further improvement after the first five years usually will come slowly and is largely dependent on weather conditions. An attempt to produce nothing but trophy-type animals is not a realistic goal. Habitat conditions and existing genetic characteristics will usually indicate the type of animals that can be produced. Vegetation may be slow to recover, and genetic improvement takes time. With time, however, quality will improve.

One of the most difficult requirements of a co-op is to collect harvest data and maintain accurate records. Some landowners and hunters may not want other members to know the details of their operations. However, these records are needed to accurately measure the progress of the association. Records should be kept about the sex, age, weight, and antler measurements of harvested animals. There are proper techniques one must know to take these measurements, and this training could be provided during one of the association meetings.

One of the main benefits of managing large tracts of land cooperatively is that a real difference can be made in the health of a large wildlife population and of the habitat that supports it. There can be few greater goals for wise land use.

Survey of State Wildlife
Agency Efforts to Encourage
Wildlife Management
on Private Lands

In 1994 Delwin Benson surveyed the chief administrators of all state wildlife agencies in the United States to learn what programs were in place to encourage wildlife management by private landowners and the granting of access to recreationists. Responses were received from all fifty states.

Ninety-six percent of state wildlife administrators believed that access to private lands was important for their organizational objectives, but 45 percent said that access for hunting had decreased while 69 percent believed leasing of lands for hunting had increased over the past ten years. Sixty-six percent said that demands to view wildlife on private lands were minor. As for habitat management on private lands leased for hunting, 39 percent said it had increased, 39 percent said it had remained the same, and 22 percent had no information. Habitat management on lands not leased was believed by 42 percent to have increased and by 42 percent to have stayed the same; 16 percent had no information.

Respondents reported that the four most important management practices landowners should implement were habitat practices (69 percent), access and hunter management (12 percent), planning and enterprise management (11 percent), and animal population management (8 percent) (table D.1). State programs that assist landowners with wildlife and habitat management for hunted and nonhunted species and for hunting management often overlap and were difficult to distinguish. Programs for hunted species (table D.2) and nonhunted species (table D.3) and a state-by-state summary of programs and budgets (table

D.5) show the array of programs and reveal some similarities. Table D.4 categorizes what wildlife agencies and private landowners must do to effect proper management of all wildlife, habitat, and users on private lands.

Table D.1. Most Important Management Practices for Landowners to Implement when Managing Wildlife, Habitat, and Hunting on Their Lands

Category	Number of Responses (N=49)
Habitat Management (total)	158
General	31
Timber	23
Upland	18
Diversity	12
Wetland	12
Riparian	9
Burning	8
Food	7
Livestock management	6
Winter	6
Water	6
Agricultural crops	5
Nesting	4
Capabilities	4
Native plants	3
Chemicals	2
Erosion	2
Access and Hunter Management (total)	27
Planning and Enterprise	
Management (total)	24
Cooperative planning	10
Liability	3
Funding	3
Laws	2
Professional help	2
Business	1
Education	1
Coordination	1
Public relations	1
Population Management (total)	19

Table D.2. Topical Programs Provided by State Wildlife Management Agencies
to Assist with Hunting, Wildlife, and Habitat Management on Private Lands

Category	Number of Responses (N=50)
Habitat	49
Technical assistance	35
Access	20
Forest management	14
Partnerships	12
Animal damage control	5
Easements	3
Property tax relief	1

Table D.3. Topical Programs Provided by State Wildlife Agencies
to Encourage Nonhunted Wildlife on Private Lands

Category	Number of Responses (N=48)
Same programs as for hunted species	25
Technical assistance	16
Species-oriented	13
Backyard wildlife	12
Nongame general	9
Watchable wildlife viewing	8
Habitat management	5
No programs	4
Forest management	1

Table D.4. Actions to Effect Proper Management of All Wildlife,
Habitat, and Users on Private Lands

Category	Number of Responses (N=48)
Cooperate and work together	24
Landowner action, rights, responsibilities, trust, and empowerment	22
Technical information and direct assistance	18
Funding	17
Educational programs	14
Ecosystems management	6
Tax relief	3
Identify and protect critical habitat	2
Simplify programs	1
Policy and regulations for land use and access	1
Liability protection	1

Table D.5.
Programs and Practices that State Wildlife Agencies Use to Assist Landowners with Hunting, Nonhunting, Wildlife, and Habitat Management on Private Lands (N=50)

Program	Brief Description	Current Budget Year
	ALABAMA	
Hunting, Wildlife, Habitat:		
Deer Management Program		$200,000
Treasure Forest (Stewardship Incentive Program)	National Stewardship Program	$75,000
Animal Damage Control		Costs vary
Nonhunted:		
Treasure Forest (Stewardship Incentive Program)	National Stewardship Program	$75,000
Nongame Wildlife Program	Activities directly aimed at nongame management. We encourage wildlife management, not game management, routinely. We provide assistance to all landowners, not just those who hunt.	$300,000
	ALASKA	
For all categories:		
None		
	ARIZONA	
Hunting and Nonhunting:		
Heritage Access Program	Access through private to public lands. Signage, sign-in/sign-out boxes—focus on consumptive and nonconsumptive activities.	$500,000
Heritage Grants-in-Aid Program	Access management and easements.	

Program	Brief Description	Current Budget Year
Heritage Access Stewardship Program	Physical access, improvements.	
Adopt-a-Ranch	Recreational groups help to manage and maintain access on a particular ranch.	

ARKANSAS

Hunting, Wildlife, Habitat:

Acres for Wildlife	Extension Program	$250,000

Nonhunted:

Blue Bird Program

Barn Owl Program

Backyard Wildlife Program

CALIFORNIA

Hunting, Wildlife, Habitat:

Private Lands Wildlife Habitat Enhancement and Management Program (PLM)	Wildlife and recreation management offering special hunting seasons, licenses and bag limits.	$90,000
Private Land Access for Upland Game Bird Hunting	Develop liability waivers, enhancements.	$125,000
California Waterfowl Habitat Program	Encourage owners of private wetlands to preserve and enhance waterfowl habitat.	$130,000
Central Valley Permanent Wetland Easement Program	Preservation and enhancement of existing and restored marshes.	$1.2 million

Nonhunted:

Watchable Wildlife	Identifies areas where wildlife can be viewed. Mostly public land but some private land included.	

Table D.5.*(continued)*

Program	Brief Description	Current Budget Year
PLM Program	Previously described. Habitat work is directed toward all wildlife.	

Hunting, Wildlife, Habitat:

Program	Brief Description	Current Budget Year
Private Lands	Programs dealing with the State Land Board, landowner recognition programs, liaison with agricultural organizations on a state and federal level, and the Forest Stewardship Program.	$105,000 for operations, additional money for leasing
CHIPS	A cooperative habitat improvement program to assist landowners.	$50,000 combined with following programs
Living Snow Fence	Provides cover for many wildlife species.	$50,000
PHIPS	Pheasant Habitat Improvement Program.	$200,000
Habitat Partnership Program	Cooperative conflict resolution program for rangeland forage and fence issues.	$650,000
Farmers Home Easement/Fee Title	New program.	
Partners for Wildlife	Private lands Extension specialist to help with private land habitat, mainly wetlands.	$350,000–500,000
Ranching for Wildlife	Requires habitat management activities while allowing flexibility in hunting opportunities.	

Program	Brief Description	Current Budget Year
Nonhunted:		
Watchable Wildlife	Among its other objectives, it encourages increasing opportunity for viewing wildlife on private lands.	$1.6 million
Backyard Wildlife	Encourages people in the Denver area to provide for backyard wildlife.	

CONNECTICUT

Program	Brief Description	Current Budget Year
Hunting, Wildlife, Habitat:		
Technical Assistance in Wildlife	Provides wildlife management advice and assistance to selected landowners and in conjunction with forest management.	
Nonhunted:		
Technical Assistance	Same as above, nongame objectives are included.	
Nonharvested Wildlife Program	Among many responsibilities, encourages landowners to employ specific practices such as nest box placement.	

DELAWARE

Program	Brief Description	Current Budget Year
Hunting, Wildlife, Habitat:		
Quail Habitat Improvement	Cost sharing to establish small game habitat.	$28,000
Habitat Consultation	Biologists provide free consultation to manage property for wildlife.	about $30,000
Nonhunted:		
Wildlife Consultation	Biologists work with landowners to manage nonhunted wildlife.	about $10,000

Table D.5.(*continued*)

Program	Brief Description	Current Budget Year
	FLORIDA	
Hunting, Wildlife, Habitat:		
Technical Guidance	Provide landowners technical guidance in managing wildlife populations (harvest techniques), enhancing wildlife habitat, and managing recreational use.	about $500,000
Nonhunted:		
Nongame Wildlife Program with Urban Wildlife Program		$3.75 million
	GEORGIA	
Hunting, Wildlife, Habitat:		
Extension	Develop management plans.	
Forest Stewardship	Develop management plans with federal funding for participants.	
Nonhunted:		
Extension		
Forest Stewardship		
Community Wildlife Certification	Towns and cities are certified by doing backyard habitat projects.	
	HAWAII	
Hunting, Wildlife, Habitat:		
Technical Program	Upon request, landowners are taught Best Management Practices	Not defined— estimate $12,000
Nonhunted:		
Forest Stewardship	Provides technical and financial assistance to private landowners.	$45,000

Program	Brief Description	Current Budget Year
Natural Area Partnerships	Provides state funds on a two-for-one basis with private funds for the management of private lands that are dedicated to conservation.	$200,000

IDAHO

Hunting, Wildlife, Habitat:		
Habitat Improvement Program	Upland game bird and waterfowl.	$300,000 to operate habitat
Nonhunted:		
Habitat Improvement Program	Same as above—incidental to game bird work, nongame habitat is produced.	$300,000
Nongame	Some funds available to encourage private habitat.	less than $10,000

ILLINOIS

Hunting, Wildlife, Habitat:		
Private Lands Management Program	Provides private landowners with technical assistance in developing and managing wildlife habitat. Just one of many programs under the administrative control of the Division of Wildlife Resources.	$1.2 million
Nonhunted:		
Division of Natural Heritage Avian Ecology Natural Areas Mammal Ecology Natural Heritage Database Plant Conservation	The Illinois Nature Preserves Commission and the Illinois Endangered Species Protection Board each have permanent staff who work with Natural Heritage personnel to enhance nonhunted	$2.7 million (for public and private lands)

Table D.5.*(continued)*

Program	Brief Description	Current Budget Year
Endangered and Threatened Species Conservation	wildlife on private lands, e.g., designation of land as nature preserves, registration of land as natural areas, development of management plans for these lands, helping landowners restore prairies and other critical habitat.	

INDIANA

Hunting, Wildlife, Habitat:

Program	Brief Description	Current Budget Year
Classified Wildlife Habitat Program	Provides property tax relief	Personal time (varies)
Game Bird Habitat Cost Share Project	Provides cost share funds for developing game bird habitat up to $100/acre.	$44,000
Cost Share Project	Provides up to 90 percent cost share for developing wildlife habitat or $1,000, whichever is less.	$37,400

Nonhunted:

No specific programs with obligated funding. This is approached through various printed information and technical assistance.

IOWA

Hunting, Wildlife, Habitat:

Program	Brief Description	Current Budget Year
Native Grass Planting and Farmstead Windbreaks	Costs shared.	$150,000
Technical Assistance		
Wetland Restoration		$50,000

Program	Brief Description	Current Budget
Nonhunted:		
Same as above		

KANSAS

Program	Brief Description	Current Budget
Hunting, Wildlife, Habitat:		
Wildlife Habitat Improvement Program	Addresses habitat loss.	$95,000
Walk-in Program	Fee paid to landowners who allow access. Pilot program in 1995, 4-county area.	$80,000
Nonhunted:		
Wildlife Habitat Improvement Program	Habitat development and technical assistance to enhance habitat for all wildlife, even though game animals are the target species.	
Backyard Wildlife Habitat Program	Provides technical and direct assistance to urbanites who develop nongame habitat.	$5,000
Owls		$20,000—varies according to grants

KENTUCKY

Program	Brief Description	Current Budget
Hunting, Wildlife, Habitat:		
Habitat Improvement Program	Technical guidance/cost share on implementation	$100,000 (cost share only)
Forest Stewardship/ Stewardship Incentive Program	Technical guidance/cost share on implementation	$300,000
Nonhunted:		
Habitat Improvement Program		
Forest Stewardship/Stewardship Incentive Program		

Table D.5.*(continued)*

Program	Brief Description	Current Budget Year
Backyard Habitat Program and Wild School Sites	Technical guidance/cost share on implementation.	about $950,000

LOUISIANA

Hunting, Wildlife, Habitat:

Program	Brief Description	Current Budget Year
Alligator	Wild alligator harvest, farming, and egg pickup.	$700,000
Deer Management Assistance Program	Provide harvest recommendations, antlerless tags, and data interpretation for clubs enrolled in program.	$150,000
Forest Stewardship	Management plans and cost-share assistance for private landowners interested in managing their lands for wildlife.	$65,000
Waterfowl Project/Private Lands Initiative	Promote wetlands management on private lands for waterfowl and other wetland species. Technical assistance and engineering.	$37,000
Natural Heritage/Nongame	Through TNC, enter into conservation agreements with private landowners with unique sites. Provide information through publications and presentations on management of nonhunted wildlife.	$220,000
Technical Assistance	General information to the public and private landowners on all aspects of wildlife and wildlife management. Accomplished with personal contact, publications, and stewardship plans.	$25,000

Program	Brief Description	Current Budget Year
Extension Activities	Develop resource management plans to carry out the Coastal Wetlands Planning, Protection, and Restoration Act.	$50,000
Nonhunted:		
Bald Eagle	Survey population and identify new nests. Regulate development near nests.	$32,000
Brown Pelican	Survey coastal nesting areas. Information is used for private lands habitat.	$5,000

MAINE

Program	Brief Description	Current Budget Year
Hunting, Wildlife, Habitat:		
Ask First/Education	Landowner relations project.	$60,000
Forest Stewardship Incentive Program	Landowner assistance.	
Technical Assistance for Habitat Management on Private Lands	Important habitat management.	$116,898
Nonhunted:		
Wildlife Program	Conservation of nonhunted species with emphasis on endangered and threatened species.	$517,000

MARYLAND

Program	Brief Description	Current Budget Year
Hunting, Wildlife, Habitat:		
Cooperative Wildlife Management Program	Provides hunters with access to private lands and public lands not controlled by the state.	more than $1 million (all programs)
Wildlife Habitat Incentive Program	Provide incentives to landowners to conserve, enhance, or create wildlife habitat. Encourages hunting and hunter access.	

Table D.5.*(continued)*

Program	Brief Description	Current Budget Year
Partners for Wildlife	Same as above.	
Wetland Restoration Program	Same as above.	
North American Waterfowl Management Program	Same as above.	
Upland Habitat Restoration Program		
Nonhunted:		
Wild Acres	Backyard habitat certification program that helps suburban and urban homeowners attract wildlife to their properties. This program is not used to directly promote hunting, but rather to educate people about wildlife habitat and habitat needs. A newsletter is sent to those enrolled.	$20,000 (mostly printing costs)

MASSACHUSETTS		
Hunting, Wildlife, Habitat:		
Technical Assistance	Provides biological review and recommendations for wildlife enhancement and management.	
Nonhunted:		
Endangered and threatened species and Natural Heritage Program	Inventories and develops management and restoration programs for nongame and endangered species.	$600,000

MICHIGAN		
Hunting, Wildlife, Habitat:		
Hunting Access Program	Leasing private lands for public hunting.	$350,000

Program	Brief Description	Current Budget Year
Working Together for Wildlife	Creating partnerships for planning, assistance, and cost sharing for habitat improvement for all species on private lands.	$150,000
Nonhunted: Working Together for Wildlife	See above.	
Natural Heritage Program	Provides information on nongame, threatened, and endangered species of plants and animals. Work with groups and agencies to conserve.	$60,000
Landowner Contact Program	Contact private landowners to make them aware of threatened and endangered plants which exist or could exist with management.	$30,000

MINNESOTA		
Hunting, Wildlife, Habitat: Pheasant Habitat Improvement Program	Habitat projects on both private and public lands; cost sharing with private landowners.	$605,000
Deer Habitat Improvement Program	Habitat projects on both public and private lands for deer enhancement.	$1.2 million
Waterfowl Habitat Improvement Program	Mostly public land, but can be used on wetland restorations.	$708,000
Nonhunted: No-game Program	Focus on species that are endangered, threatened, or of special concern and on watchable wildlife.	$1 million

Table D.5.*(continued)*

Program	Brief Description	Current Budget Year
	MISSISSIPPI	
Hunting, Wildlife, Habitat:		
Overall Technical Guidance Program	Federal aid project involving all game species of wildlife.	$256,000
Deer Management Assistance	Assists landowners/clubs with meeting specific deer management objectives.	$ included with technical guidance program
Beaver Control Assistance Program	Joint program with several state agencies and USDA/ADAC to control nuisance beavers.	$164,000 (dept. share)
Partners in Wildlife	Joint program with federal agencies and private organizations; directed primarily at waterfowl in the Mississippi Delta.	$25,000 (dept. share)
Nonhunted:		
Project Wild	Environmental education curriculum.	
	MISSOURI	
Hunting, Wildlife, Habitat:		
Conservation Reserve Program Incentive	Encourage selection of cover beneficial to wildlife.	$250,000
Prairie Restoration Incentive	Restore vigor to prairie that has been mismanaged.	$100,000
Wetland Reserve Program Incentive	Encourage Wetland Reserve Program sign-up by helping with restoration costs.	$100,000

Program	Brief Description	Current Budget Year
Partners for Prairie Wildlife	Restore prairie ecosystems.	$80,000
Food and Cover Program	Free or low-cost seed/trees for wildlife plantings.	$40,000

Nonhunted:
All but the food/cover program
listed above

MONTANA

Hunting, Wildlife, Habitat:

Habitat Montana	Dept. interest (title, leases, easement) on private land.	$3.5 million
Upland Bird Habitat Enhancement Program	Cooperate to develop habitat for upland birds on private land.	$900,000
Waterfowl Program	Enhancements and improvements for waterfowl production on private lands.	$500,000
Block Management Program	Hunter access and management on private lands.	$425,000

Nonhunted:

Watchable Wildlife Program	Encourages appreciation and opportunities for nongame wildlife viewing. Has mainly public land focus.	$55,400

NEBRASKA

Hunting, Wildlife, Habitat:

Wetlands Initiative Program	Provides funding for wetland restoration, enhancement, or creation. Also pays landowners' costs for restoration under the Wetland Reserve Program (WRP).	$100,000

Table D.5.*(continued)*

Program	Brief Description	Current Budget Year
	Waterbank bonus/incentive practice. Pays a one-time cash bonus to landowners enrolling wetlands in the WRP.	$100,000
Wildlife Habitat Improvement Program	Cooperative with natural resources districts to create new or enhanced wildlife habitat. Cost-share rates: NG&PC 75 percent, NRD 25 percent.	$560,000
Shelterbelt Program	Program to establish shelter belts.	$40,000
Tree and Shrub Program	Provides free tree and shrub seedlings. 100 trees and shrubs per packet.	$20,000
Food Plot Program	Provides free food plot seed mixtures.	$6,000
Wildlife Depredations	Provides technical assistance and materials for depredation problems.	Based on need
Information & Education	Publications, workshops, etc., related to species, habitat, and hunting.	Varies
Technical Assistance	Personnel provide technical assistance directly to landowners and indirectly through other government agencies such as USDA.	Varies
Nonhunted: Programs listed above are also targeted to nonhunted wildlife.		
Special Publications	Target threatened and endangered species, bird houses, bird feeders, etc.	Varies

Program	Brief Description	Current Budget Year

NEVADA

Hunting, Wildlife, Habitat:
No specific programs in place because of the small amount of private land

Most of our resources are focused on public land (87 percent of the state). We do have a deer and antelope compensation tag program to issue tags to landowners who experience depredation from deer or antelope. One tag is issued for each 50 deer or antelope doing damage to private lands.

Nonhunted:
Technical Assistance

Assistance provided to private landowners.

NEW HAMPSHIRE

Hunting, Wildlife, Habitat:		
Extension Biologist		$50,000
Cooperative Sign Program		$1,000
Regional Biologist	Responds to questions.	$10,000
Nonhunted:		
Nongame Biologist	Inventory and education.	$100,000
Wildlife Education	Inventory and education.	$70,000

NEW JERSEY

Hunting, Wildlife, Habitat:		
Technical advice on establishing hunting program	Assist in determining hunter density, harvest quotas, etc.	Relatively little cost
Technical advice on habitat improvement	General information on food and cover needs.	Relatively little cost

Table D.5.*(continued)*

Program	Brief Description	Current Budget Year
Wildlife Damage Control	Provide fencing, repellent, and permits to shoot depredating deer.	$200,000
Nonhunted: Watchable Wildlife	Develop areas for different species on private and public lands.	about $600,000
Nest Box Program	Encourage nesting.	about $10,000
Osprey Program		

NEW MEXICO

Program	Brief Description	Current Budget Year
Hunting, Wildlife, Habitat: Landowner-Sportsman Cooperative Program	Improve relations between landowner/sportsman/agency	$200,000
Private Lands Habitat Specialist	Advise/assist private landowners to improve habitat.	$50,000
Nonhunted: Watchable Wildlife Program	76 sites marked along roadways to allow public viewing.	
Private Lands Habitat Specialist	Advise/assist private landowners.	$50,000

NEW YORK

Program	Brief Description	Current Budget Year
Hunting, Wildlife, Habitat: "Ask" Program	Landowners post signs advising that permission may be given for access.	
Nonhunted: Wildlife Observation	Program in development.	

Program	Brief Description	Current Budget Year

NORTH CAROLINA

Hunting, Wildlife, Habitat:

Forest Stewardship Program	Write management plans for landowners.	$250,000
Technical Guidance	Meet with landowners and offer advice.	$100,000
Deer Management Assistance Program	Manage deer herds at clubs' request.	$150,000

Nonhunted:
None

NORTH DAKOTA

Hunting, Wildlife, Habitat:

| Private Land Initiative | Habitat development on private land, nesting habitat, tree planting, food plots, etc. | $900,000 |

Nonhunted:

| Nongame Program | Feeders, bluebird houses, outdoor wildlife learning centers. | $30,000 |

OHIO

Hunting, Wildlife, Habitat:

Technical Assistance	Direct technical assistance to private landowners, habitat-oriented, not just huntable species.	9 biologists—about $300,000
Wetland Cost-Sharing	Reimbursement for restoration.	$75,000
Hunting with Permission	Assist in managing hunting pressure.	$25,000
Wildlife Food Plots	Free seed.	$10,000

Nonhunted:

| Technical Assistance | See above. | |

Table D.5.*(continued)*

Program	Brief Description	Current Budget Year

OKLAHOMA

Hunting, Wildlife, Habitat:

Deer Management Assistance	Provides cooperators with information necessary to implement more intensive deer management.	$11,240
Quail Enhancement	Research on bobwhite biology, requirements, and management. Technical assistance to improve habitat.	$90,900
Forest Stewardship	Assist rural landowners to implement multiple-use management strategy.	$26,000
Wildlife Habitat Improvement Program	Provides cooperators with technical assistance and matching funds for wildlife habitat enhancement primarily for deer, quail, doves, pheasants, and waterfowl.	$50,000

Nonhunting:

Nongame Wildlife	Funded primarily by donations from state income tax refund. Production of nongame brochures, books, posters, slide shows, and newsletters. Cooperation with other agencies to reintroduce bald eagles, river otters, and prairie dogs. Creation of watchable wildlife opportunities. Direct protection and research of endangered species and other species of concern. Public information about the Nongame Program.	$232,000

Program	Brief Description	Current Budget Year

OREGON

Hunting, Wildlife, Habitat:
Habitat Improvement Program

Wildlife Habitat on Public
and Private Lands

Damage Control Program	Help landowners to prevent and correct wildlife damage.	
Green Forage Program	Assist landowners experiencing crop damage.	
D.E.A.R. Program	Deer Enhancement and Restoration for mule deer habitat on private lands.	
Landowner Preference Program	Landowners get tag for private land if unsuccessful in drawing.	
Access and Habitat Board Program	Increase hunting access to private lands.	$500,000
Upland Bird Stamp Program	Access, habitat, etc.	$300,000

Nonhunted:

Song Bird Nest Boxes/Bat Houses	Provides materials and brochures on how to build houses and attract wildlife.	$5,000
Volunteer Wildlife Surveys	Conduct surveys on private lands to understand nongame species needs and develop appreciation of wildlife by nonhunters.	$20,000
Naturescaping	Turning backyards or small land holdings into attractive wildlife sites through habitat management, etc. Complete book for sale!	$2,000

Table D.5.*(continued)*

Program	Brief Description	Current Budget Year

PENNSYLVANIA

Hunting, Wildlife, Habitat:

Public Access Programs		about $900,000

Nonhunted:

Kestrel, Blue Bird, Bat Boxes	For distribution to cooperators.	
Wildlife Food Plot Seed Mix		$22,000 (to purchase)
Species Restorations	Otter, peregrine falcon, osprey, fisher, etc.	
Gating Bat Hibernacula	To limit human access.	$4,000

RHODE ISLAND

Hunting, Wildlife, Habitat:

Landowner Coop. Project	Sign, seeds, fertilizer.	

Nonhunted:
None

SOUTH CAROLINA

Hunting, Wildlife, Habitat:

Technical Assistance (regional staff of 18 biologists)	Provide management advice and written recommendations to landowners—approx. 600 plans/year.	Handled through our normal PR program at the regional level
Statewide Projects (8 biologists)	Technical assistance from project leaders on specific species, e.g., deer, quail, etc.	Financed through license revenue

Program	Brief Description	Current Budget Year
Forest Stewardship (2 full-time biologists)	Wildlife assistance to private landowners—200 per year.	$96,000
Nonhunted: Wildlife Diversity Section	Technical assistance on nongame.	
South Carolina Wildlife Magazine	Article to encourage management for nongame.	

SOUTH DAKOTA

Program	Brief Description	Current Budget Year
Hunting, Wildlife, Habitat: Pheasants for Everyone	Private land management program.	
Wildlife Habitat Improvement Practices	Tree, grass, and food plantings on private lands.	$460,000
Walk-in Areas	Private land hunting access program.	$300,000
March & Habitat USA	Cooperative wetland management and creation program with Ducks Unlimited.	$120,000
Nonhunted: Although not specifically defined in the program, both WHIP and our wetland program were designed to benefit a wide array of game and nongame species	For example, tree plants are beneficial to many unhunted species of neotropical migrant birds.	

TENNESSEE

Program	Brief Description	Current Budget Year
Hunting, Wildlife, Habitat: Upland Game Bird Habitat Program	Small-game habitat technical assistance and cost-sharing.	$135,000
Technical Assistance	Habitat evaluation/management plans for most species.	Funding for some available

Table D.5.*(continued)*

Program	Brief Description	Current Budget Year
Forest Stewardship/ Stewardship Incentive Program	Cooperation provided; program administered through Division of Forestry.	
Free Wood Duck Boxes		
Nonhunted:		
Wildlife Observation Areas	Landowner agreement with state to provide public access and sometimes enhancement of wildlife viewing opportunities.	$500
Urban Wildlife	Technical advice, primarily through publications provided for attracting desired wildlife and discouraging undesired wildlife.	$25,000
Free Blue Bird Boxes		

TEXAS		
Hunting, Wildlife, Habitat:		
Private Lands Enhancement Program	Providing technical guidance and planning for habitat enhancement.	
"Making Tracts for Texas Wildlife" Newsletter	A communications tool for private landowners, carrying information about habitat enhancement techniques.	
Public Hunting Program	Leasing private lands for public hunting.	
Hunters Clearinghouse Directory	A listing of private hunting leases.	
Nonhunted:		
Private Lands Enhancement Program	Providing technical guidance and planning for habitat enhancement.	

Program	Brief Description	Current Budget Year
Texas Private Lands Initiative	Providing technical guidance and funding assistance.	$300,000
"Making Tracts for Texas Wildlife" Newsletter	Communications to private landowners about habitat enhancement.	$20,000
Texas Wildscapes Program	Providing information and support for backyard habitat development.	$60,000
Nongame Newsletter	Providing information about nongame recreation/habitat enhancement.	$60,000

UTAH

Hunting, Wildlife, Habitat:

Posted Hunting Unit Program	Upland and big-game harvest management coop. programs for access to public in exchange for selected private hunts.	about $40,000 for big and small game together (funded by high-bid permits, application fees, and general fund)
Big Game Habitat Development	Coop. development of wildlife habitat on ranges including Conservation Reserve Program, fire-damaged shrub lands, pinion and juniper conversions, etc. Support includes consultation, planning, seed/seedlings, and application cost.	$850,000 (for all habitat programs, including cooperative programs with state and federal agencies and private lands) with $700,000 committed to programs, e.g., seed and application, etc.

Nonhunted:

Habitat Development (see second item)	Habitat development objectives include rangeland improvement,	$850,000 (for all habitat programs,

Table D.5.(continued)

Program	Brief Description	Current Budget Year
	watershed maintenance objectives that provide for improved riparian systems, habitat diversity, increased productivity, etc.	including cooperative programs with state and federal agencies and private lands)

VERMONT

Hunting, Wildlife, Habitat:

On-site Technical Assistance	Review property with landowners and land managers and help develop management plants.	Not currently available.
Information and Education	Provide training materials for landowners and land managers through guideline books, workshops, etc.	Not currently available.
Phone Consultation		

Nonhunted:

| Backyard Wildlife Habitat | Slide show and educational booklet. | Not currently available. |

VIRGINIA

Hunting, Wildlife, Habitat:

| Deer Management Assistance Program | Objective is to allow landowner and hunt club to manage deer populations on their property. | $103,993 |
| Forest Stewardship | Provide management assistance to small landowners. | $92,410 |

Nonhunted:

| Forest Stewardship | See above | |

WASHINGTON

Hunting, Wildlife, Habitat:

| Private Lands Wildlife Management Areas | Pilot program that encourages larger (>5,000 acres) private landowners to make significant | $8,500 staff $2,500 aerial flight time |

Program	Brief Description	Current Budget Year
	habitat improvements in exchange for more control over hunting seasons and permit levels. Wildlife viewing, photography, and other nonhunting recreation are also encouraged. Habitat improvements that specifically target nongame wildlife are also required as part of the program, especially for rare species. Currently only two landowners are participating.	$500 travel costs
Upland Wildlife Restoration Program	Provides technical advice and materials for habitat improvements on private lands. This program is closely coordinated with the Natural Resources Conservation Service and other U.S. Department of Agriculture programs. Another segment of this program is designed to provide hunter access to private land by providing signs, reduced landowner liability, and a higher level of enforcement presence. There are currently about 1 million acres under the access program and more than 300 landowners under either habitat or access agreements.	$1 million
Nonhunted:		
Backyard Wildlife Sanctuary	Provides technical information, signs, and certificates for private lands that emphasize urban backyard landscaping for wildlife. Also works with planners and developers to incorporate wildlife needs in community and housing developments.	more than $150,000

Table D.5.*(continued)*

Program	Brief Description	Current Budget Year

WEST VIRGINIA

Hunting, Wildlife, Habitat:

No specific program	Technical assistance is available to any landowner on request. Mostly advice and suggestions but can go as far as actually preparing a management plan.	No specific budget

Nonhunted:

Nongame Wildlife Program	On a limited basis, educational/informational materials on wildlife plantings, feeders, houses, etc., are distributed to those who request this information.	n/a

WISCONSIN

Hunting, Wildlife, Habitat:

Wetland Restoration		$250,000
Grassland Restoration		$75,000

Nonhunted:

Wetland Restoration		$250,000
Grassland Restoration		$75,000
U.S. Forest Service Stewardship Incentive Program		$100,000

WYOMING

Hunting, Wildlife, Habitat:

Habitat Extension Program	Cooperative agreement with Natural Resources	$127,344

Program	Brief Description	Current Budget Year
	Conservation Service to provide technical expertise (labor and budgets) to private landowners for wildlife habitat improvements.	

Nonhunted:

None

References

Acker, G. Elaine. 1995. "Outdoor Heritage Series." *Texas Parks and Wildlife* February:48–49.

Adams, C. E., and J. K. Thomas. 1986. "Characteristics and Opinions of Texas Hunters." *Proceedings of the 1986 International Ranchers Roundup.* Kerrville, Texas. 255–61.

Adams, J. S., and T. O. McShane. 1992. *The Myth of Wild Africa: Conservation Without Illusion.* New York: W. W. Norton & Company, Inc.

Agee, J. K., and D. R. Johnson. 1988. *Ecosystem Management for Parks and Wilderness.* Seattle: University of Washington Press.

Altiere, M. A. 1990. "How Common Is Our Common Future?" *Conservation Biology* 4:102–103.

Applegate, J. E. 1981. "Landowners' Behavior in Dealing with Wildlife Values." In *Wildlife Management on Private Lands,* ed. R. T. Dumke, G. V. Burger, and J. R. March, 64–72. Madison, Wis.: Department of Natural Resources.

———. 1989. "Patterns of Early Desertion Among New Jersey Hunters." *Wildlife Society Bulletin* 17:476–81.

Arha, K. 1996. "Sustaining Wildlife Values on Private Lands: A Survey of State Programs for Wildlife Management on Private Lands in California, Colorado, Montana, New Mexico, Oregon, Utah, and Washington." *Transactions of the North American Wildlife and Natural Resources Conference,* 61:267–73

"Arizona Public Land Policy." 1990. *Hunting Ranch Business* 5(9):4.

Associated Press. 1997. "Landowners To Be Paid for Conservation." *Bryan–College Station Eagle,* 17 February:A4.

Barclay, J. S. 1966. Significant factors influencing the availability of privately owned rural land to the hunters. Master's thesis, Pennsylvania State University, University Park.

Barkley, P. W., and D. W. Seckler. 1972. *Economic Growth and Environmental Decay. The Solution Becomes the Problem.* New York: Harcourt Brace Jovanovich, Inc.

Barnes, J. I., and M. C. Kalikawe. 1994. "Game Ranching in Botswana: Constraints and Prospects." In *Wildlife Ranching: A Celebration of Diversity,* ed. W. van Hoven, H. Ebedes, and A. Conroy, 245–52. Pretoria, Republic of South Africa: Promedia.

Batie, S. S., and D. B. Taylor. 1990. "Cropland and Soil Sustainability." In *Natural Resources for the 21st Century,* ed. R. N. Sampson and D. Hain, 56–77. Washington, D.C.: American Forestry Association, Island Press.

Bean, M. J. 1983. *The Evolution of National Wildlife Law.* New York: Praeger Publishing.

Benson, D. E. 1976. "Needed: Landowner Respect." *Colorado Outdoors* 25(5):9–12.

———. 1986. "Wildlife for Profit: Can Colorado Producers Learn from South Africa?" *Rancher and Farmer* August:26.

———. 1987a. "Holistic Ranch Management and the Ecosystem Approach to Wildlife Conservation." In *Proceedings, Privatization of Wildlife and Public Lands Access Symposium,* 67–69. Wyoming Game and Fish Department. Casper, Wyoming.

———. 1987b. "How To Get Started." In *Proceedings, Lease Hunting: Pros and Cons Telenet Conference*, 11–12. Kansas State University Cooperative Extension Service. Manhattan, Kansas.

———. 1987c. "Paying To Hunt on Private Lands." *Colorado Wildlife* May/June:13.

———. 1987d. "Planning for Wildlife and Hunting on Private Land in the West." In *Proceedings, Landowner/Sportsman Relations Conference*, 75–79. Colorado Division of Wildlife. Denver, Colorado.

———. 1988a. "Education to Promote Wildlife Values on Private Land." *Human Dimensions in Wildlife Newsletter* 7:27–29.

———. 1988b. "Integrating Public and Private Concerns for Managing Wildlife and Hunters." In *Proceedings, Recreation on Rangelands: Promise, Problems, Projections*. Society for Range Management, 77–81. Corpus Christi, Texas.

———. 1988c. "What Fee Hunting Means to Sportsmen in the U.S.A.: A Preliminary Analysis." In *Proceedings of the 1st International Wildlife Ranching Symposium*, 296–302. Las Cruces, New Mexico.

———. 1989a. "Changes from Free to Fee Hunting." *Rangelands* 11(4):176–80.

———. 1989b. Private values and management of wildlife and recreation in South Africa with comparisons to the U.S.A. Ph.D. diss., Colorado State University, Fort Collins.

———. 1989c. "What Fee Hunting Means to Sportsmen." In *Proceedings of the 1st International Wildlife Ranching Symposium*. Las Cruces, New Mexico.

———. 1990. "Wilderness Values on Western Ranches." In *The Use of Wilderness for Personal Growth, Therapy and Education*, ed. A. T. Easley et al., 130–34. USDA Forest Service General Technical Report RM 193. Rocky Mountain Forest and Range Experiment Station. Fort Collins, Colorado.

———. 1991a. "Integrated Wildlife Management Partnerships Among Agriculture, Natural Resources Professions and Business." In *Wildlife Production: Conservation and Sustainable Development*, ed. L. A. Renecker and R. J. Hudson, 344–48. Fairbanks: University of Alaska.

———. 1991b. "Values and Management of Wildlife and Recreation on Private Land in South Africa." *Wildlife Society Bulletin* 19:497–510.

———. 1991c. *Wildlife and Recreation Enterprise Directory: Colorado Private Recreation Lands*. Colorado State University Cooperative Extension Service XCM-153. Fort Collins, Colorado.

———. 1992. "Commercialization of Wildlife: A Value-added Incentive for Conservation." In *The Biology of Deer*, ed. R. D. Brown, 539–53. New York: Springer-Verlag.

———. 1993. "An Agenda for Conservation and Resource Management." In *Conservation and Resource Management*, ed. S. K. Majumdor, E. W. Miller, D. E. Baker, E. K. Brown, J. R. Pratt, and R. F. Schmalz, 424–35. Easton, Pa.: Academy of Science.

Berry, W. 1989. "The Futility of Global Thinking." *Harpers* September:53–54.

Berryman, J. H. 1981. "Needed Now: An Action Program to Maintain and Manage Wildlife Habitat on Private Lands." In *Wildlife Management on Private Lands*, ed. R. T. Dumke, G. V. Burger, and J. R. March, 6–10. Madison, Wis.: Department of Natural Resources.

———. 1983. "Wildlife Damage Control: A Current Perspective." In *Proceedings of the 1st Eastern Wildlife Damage Control Conference*, ed. D. J. Decker, 3–5. Ithaca, N.Y.: Cooperative Extension Service.

———. 1987. "Socioeconomic Values of Wildlife Resources: Are We Really Serious?" In *Valuing Wildlife—Economic and Social Perspectives*, ed. D. J. Decker and G. R. Goff, 5–11. Boulder, Colo.: Westview Press.

Binger, C. R. 1975. "Corporate Views and Responsibilities for Public Values and Profits." *Transactions, North American Wildlife and Natural Resources Conference* 49:405–12.

Bolin, E. G. 1989. "Conservation Biology, Wildlife Management, and Spaceship Earth." *Wildlife Society Bulletin.* 17:351–54.

Bothma, J. Du P., and M. A. Rabie. 1983. "Wild Animals." In *Environmental Concerns in South Africa—Technical and Legal Perspectives*, ed. R. F. Fuggle and M. A. Rabie, 190–236. Cape Town, Republic of South Africa: Juta & Co. Ltd.

Bovard, J. 1995. *Lost Rights: The Destruction of American Liberty.* New York: St. Martin's Griffin.

Bromley, P. T., and D. E. Benson. 1987. *Supplemental Income from Wildlife on Your Land.* U.S. Department of Agriculture Cooperative Extension Service.

Brown, P. J., and G. E. Haas. 1980. "Wilderness Recreation Experience: The Rawah Case." *Journal of Leisure Research* 12(3):229–41.

Brown, T. L. 1974. "New York Landowners' Attitudes toward Recreation Activities." *Transactions, North American Wildlife and Natural Resources Conference* 39:173–79.

Brown, T. L., J. Decker, and J. W. Kelley. 1984. "Access to Private Lands for Hunting in New York: 1963–1980." *Wildlife Society Bulletin* 12:344–49.

Brown, T. L., D. J. Decker, K. G. Purdy, and G. F. Mattfeld. 1987. "The Future of Hunting in New York." *Transactions, North American Wildlife and Natural Resources Conference* 53:553–66.

Bubenik, A. B. 1976. "Evolution of Wildlife Harvesting Systems in Europe." *Transactions, Fed.-Prov. Wildlife Conference* 40:97–105.

———. 1989. "Sport Hunting in Continental Europe." In *Wildlife Production Systems*, ed. R. J. Hudson, K. R. Drew, and L. M. Baskin, 115–33. Cambridge, New York, Port Chester, Melbourne, and Sydney: Cambridge University Press.

Burger, G. V., and J. G. Teer. 1981. "Economic and Socioeconomic Issues Influencing Wildlife Management on Private Land." In *Wildlife Management on Private Lands*, ed. R. T. Dumke, G. V. Burger, and J. R. March, 252–78. Madison, Wis.: Department of Natural Resources.

Campbell, A. 1994. *Landcare: Communities Shaping the Land and the Future.* Saint Leonards, New South Wales, Australia: Allen & Unwin Pty. Ltd.

Caro, T. J. 1986. "The Many Paths to Wildlife Conservation in Africa." *Oryx* 20(4):221–29.

Carson, R. 1962. *Silent Spring.* Cambridge, Mass.: Riverside Press.

Catton, R. C., Jr. 1969. "Motivations of Wilderness Users." *Pulp and Paper Magazine of Canada* 121–26.

Chamberlain, P. A. 1984. Waterfowl and agriculture—an assessment of wintering waterfowl management and land-use economics on the Texas High Plains. Ph.D. diss., Texas Tech. University, Lubbock.

Chandler, W. J., ed. 1988. *Audubon Wildlife Report 1988–1989.* New York: National Audubon Society.

Child, G. 1980. "The Future of Wildlife in Zimbabwe, Both Inside and Outside the Parks and Wildlife Estate." *Zimbabwe Wildlife* 3:4–6.

———. 1995. *Wildlife and People: The Zimbabwean Success.* Harare, Zimbabwe, and New York: Wisdom Foundation.

Ciriacy-Wantrup, S. V. 1968. *Resource Economics and Policies*. 3rd edition. University of California Agricultural Experiment Station. Berkeley, California.

Clark, R. N., J. C. Hendee, and F. L. Campbell. 1971. "Values, Behavior and Conflict in Modern Camping Culture." *Journal of Leisure Research* 3(3):143–59.

Cole, D. N. 1990. "Wilderness Management: Has It Come of Age?" *Journal of Soil and Water Conservation* 45:360–64.

Colorado Division of Wildlife. 1972. Summary of guides and outfitters and clients, 1970 and 1971. Unpublished report.

Colvin, I. S. 1983. An inquiry into game farming in the Cape Province. Master's thesis, University of Cape Town, South Africa.

Cordell, H. K. 1990. "Outdoor Recreation and Wilderness." In *Natural Resources for the 21st Century*, ed. R. N. Sampson and D. Hain, 242–68. Washington, D.C.: American Forestry Association, Island Press.

Cordell, H. K., et al. 1988a. *An Analysis of the Outdoor Recreation and Wilderness Situation in the United States: 1989–2040*. A technical document supporting the 1989 RPA assessment. USDA Outdoor Recreation and Wilderness Assessment Research Unit, Southeastern Forest Experiment Station.

Cordell, H. K., B. Wright, A. Powell, R. Guldin, and F. Kaiser. 1988b. Leasing values and management of wildlife and fish resources on private lands in the United States. Paper presented at the 53rd North American Wildlife and Natural Resources Conference, Louisville, Kentucky.

Council on Environmental Quality. 1977. *The Evolution of National Wildlife Laws*. Washington, D.C.: U.S. Government Printing Office.

Cox, G. W., ed. 1969. *Readings in Conservation Ecology*. New York: Appleton-Century-Crofts.

Cronquist, M. V. 1990. Attitudes of pronghorn hunters for paying to hunt on private lands in Colorado. Master's thesis, Colorado State University, Fort Collins.

Crook, B. J. S. 1997. Southern African community-based natural resource use programs: Factors contributing to success. Master's thesis, Colorado State University, Fort Collins.

Davies, R. J., D. Grossman, and L. Rammutla. 1994. "Wildlife Use and Community Development in Bophuthatswana." In *Wildlife Ranching: A Celebration of Diversity*, ed. W. van Hoven, H. Ebedes, and A. Conroy, 258–59. Pretoria, Republic of South Africa: Promedia.

Davis, R. K. 1964. "The Value of Big Game Hunting in a Private Forest." *Transactions, North American Wildlife and Natural Resources Conference* 29:393–403.

———. 1985. "Research Accomplishments and Prospects in Wildlife Economics." *Transactions, North American Wildlife and Natural Resources Conference* 50:392–404.

———. 1995. "A New Paradigm for Wildlife Conservation: Using Markets to Produce Big Game Hunting." In *Wildlife in the Marketplace*, ed. T. L. Anderson and P. J. Hill, 109–25. Lanham, Md.: Rowman and Littlefield Publishing, Inc.

Davis, R. K., and D. E. Benson. 1994. "An Evaluation of Colorado's Ranching Program for Indigenous Wildlife." In *Wildlife Ranching: A Celebration of Diversity*, ed. W. van Hoven, H. Ebedes, and A. Conroy, 274–81. Pretoria, Republic of South Africa: Promedia.

Decker, E. 1987. Institutional and financial considerations in wildlife management. International Conference on Wildlife Management. United National FAO-IUCN, 1–5.

Decker, E., and J. Nagy. 1989. "European Wildlife Management." In *Proceedings, First International Wildlife Ranching Symposium,* ed. R. Valdez, 20–23. Las Cruces, New Mexico.

Doig, H. E. 1990. "Natural Resources Management Needs for Private Landowners in the Northeast." In *R. D. No. 740,* ed. W. N. Grafton, A. Ferrise, D. Colyer, D. K. Smith, and J. E. Miller, 17–24. West Virginia University Extension Service. Morgantown, West Virginia.

Downing, K., and R. N. Clark. 1979. "Users' and Managers' Perceptions of Dispersed Recreation Impacts: A Focus on Roaded Lands." In *Proceedings, Recreational Impact on Wildlands Conference,* 18–23. Seattle, Washington.

Drury, M. 1982. "Wildlife in Marginal Areas." *Zimbabwe Wildlife* 29:24–27.

Dumke, R. T., G. V. Burger, and J. R. March. 1981. *Wildlife Management and Private Lands.* La Crosse, Wis.: La Crosse Printing Co., Inc.

Du Toit, R. F. 1994. "Large-scale Wildlife Conservancies in Zimbabwe: Opportunities for Commercial Conservation of Endangered Species." In *Wildlife Ranching: A Celebration of Diversity,* ed. W. van Hoven, H. Ebedes, and A. Conroy, 195–300. Pretoria, Republic of South Africa: Promedia.

Dyer, M. I., and M. M. Holland. 1988. "UNSCO's Man and the Biosphere Program." *BioScience* 38:635–41.

Easterbrook, G. 1995. *A Moment on the Earth: The Coming Age of Environmental Optimism.* New York: Penguin Books, Ltd.

Edwards, C. E. 1989. "The Wildlife Society and the Society for Conservation Biology: Strange But Unwilling Bedfellows." *Wildlife Society Bulletin* 17:340–43.

Edwards, S. R. 1995. "Conserving Biodiversity: Resources for our Future." In *The True State of the Planet,* ed. R. Bailey, 211–67. New York: Free Press.

Eltringham, S. K. 1984. *Wildlife Resources and Economic Development.* Chichester, N.Y., Brisbane, Toronto, and Singapore: John Wiley and Sons.

Fambrough, W. J. 1987. *The Texas Deer Lease.* Publication No. 570. Real Estate Center. Texas A&M University, College Station, Texas.

Fazio, J. R., and D. L. Gilbert. 1981. *Public Relations and Communications for Natural Resource Managers.* Dubuque, Iowa: Kendall/Hunt Publishing Company.

Feltner, G. 1972. "Non-resident Hunting Down Almost in Half in '71." *Colorado Outdoors* July–August:22.

Finance Week. 1990. "Buck in the Bag SA is a Number One Big Game Hunting Destination." *FOCUS on South Africa* March:15.

Fitzhugh, E. L. 1989. "Innovation of the Private Land Wildlife Management Program: A History of Fee Hunting in California." *Transactions, Western Section, Wildlife Society* 25:49–59.

Forrest, N. K. 1968. Effects of commercialized deer hunting arrangements on ranch organization, management, costs, and income: The Llano Basin of Texas. Master's thesis, Texas A&M University, College Station.

Freese, C. 1994. *The Commercial, Consumptive Use of Wild Species: Implications for Biodiversity Conservation.* International Project 9Z0534.04. Washington, D.C.: World Wildlife Fund.

Frerich, S. J., J. W. Mjelde, J. R. Stoll, and R. C. Griffin. 1989. Institutions concerning wildlife on privately-owned land: Results from a survey of State Wildlife Departments. Unpublished paper, Dir 889-1, SP-2, Department of Agricultural Economics, Texas A&M University, College Station.

Fuggle, R. F. 1983. "Nature and Ethics of Environmental Concerns." In *Environmental Concerns in South Africa: Technical and Legal Perspectives*, ed. R. F. Fuggle and M. A. Rabie, 1–8. Cape Town, Republic of South Africa: Juta & Co. Ltd.

"The Future of Wildlife Resources Cannot Afford Strange or Unwilling Bedfellows." 1989. *Wildlife Society Bulletin* 17:343–44.

Gardner, A. 1998. "White-tailed Deer." In *Producing Quality Whitetails*, ed. A. Brothers and M. E. Ray Jr., 20–32. Laredo, Tex.: Wildlife Services, 1975. Reprint, Kingsville, Tex.: Caesar Kleberg Wildlife Research Institute, Texas A&I University.

Gasset, J. O. 1972. *Meditations on Hunting*. New York: Charles Scribner's Sons.

Geist, V. 1985. "Game Ranching: Threat to Wildlife Conservation in North America." *Wildlife Society Bulletin* 13:594–98.

———. 1986. "Wildlife a Public Trust." *Western Sportsman* July/August:52–55.

———. 1987. "Three Threats to Wildlife: Game Markets, Pay Hunting and Hunting for 'Fun.'" In *Proceedings, Privatization of Wildlife and Public Land Access Symposium*, 46–58. Cheyenne: Wyoming Game and Fish Department.

———. 1988. "How Markets in Wildlife Meat and Parts, and the Sale of Hunting Privileges, Jeopardize Wildlife Conservation." *Conservation Biology* 2(1):1–12.

———. 1989. "Legal Trafficking and Paid Hunting Threaten Conservation." *Transactions, 54th North American Wildlife and Natural Resources Conference* 54:171–78.

Gilbert, A. H. 1977. "Influence of Hunter Attitudes and Characteristics on Wildlife Management." *Transactions, North American Wildllife and Natural Resources Conference* 42:227–35.

Gilbert, F. F., and D. G. Dodds. 1992. *The Philosophy and Practice of Wildlife Management*. Malabar, Fla.: Robert E. Krieger Publishing Co.

Gonzales, S. R. 1989. "Access Systems for Private Lands in New Mexico." *Transactions, North American Wildlife and Natural Resources Conference* 54:179–82.

Gordon, R. E. 1995. *National Wilderness Institute Resource* 6(1).

Guynn, D. E., and J. L. Schmidt. 1984. "Managing Deer Hunters on Private Lands in Colorado." *Wildlife Society Bulletin* 12:12–19.

Hagenstein, P. 1990. "Forests." In *Natural Resources for the 21st Century*, ed. R. N. Sampson and D. Hain, 78–100. Washington, D.C.: American Forestry Association, Island Press.

Halse, A. R. D. 1983. "A Look at Game Farming in the Eastern Cape." *Pelea* 2:85–89.

Hamilton, P. R. 1978a. "Recreation, Menace or Milestone?" *Cattleman* June:42, 68, 70.

Hamilton, P. R. 1978b. "Recreational Estate in Texas." *Texas Bar Journal* June:511–16.

Hanks, F., W. D. Densham, G. L. Smuts, J. F. Jooste, S. C. J. Joubert, P. Leroux, and P. Les Milstein. 1981. "Management of Locally Abundant Mammals—the South African Experience." In *Problems in Management of Locally Abundant Wildlife Mammals*, 21–55. South Africa: Academic Press, Inc.

Hardin, G. 1968. "The Tragedy of the Commons." *Science* 162:1243–48.

Hawley, A. W. L., ed. 1993. *Commercialization and Wildlife Management: Dancing with the Devil*. Malabar, Fla.: Krieger Publishing Co.

Hendee, J. C. 1974. "A Multi-satisfaction Approach to Game Management." *Wildlife Society Bulletin* 2(3):104–13.

Hendee, J. C., and R. W. Harris. 1970. "Foresters' Perception of Wilderness—Users' Attitudes and Preferences." *Journal of Forestry* 68(12):759–62.

Hey, D. 1977. "The History and Status of Nature Conservation in South Africa." In *A History of ScientiWc Endeavour in South Africa,* ed. A. C. Brown, 132–63. Wynberg Cape, Republic of South Africa: The Rustica Press (PTY.) Ltd., Royal Society of South Africa.

Hill, J. 1976. "Texas Trespass Laws." *Cattleman* May:34, 114, 116, 118.

Holecek, D. F., and R. D. Westfall. 1977. *Public Recreation on Private Lands—the Landowners' Perspective.* Michigan Agricultural Experiment Station Research Report 335.

Horton, C. L. C. 1992. An interpretive analysis of Texas ranchers' perceptions of endangered species management policy as related to management of golden-cheeked warbler habitat. Master's thesis, Texas A&M University, College Station.

Howard, P. C., and A. N. Marchant. 1984. "The Distribution and Status of Some Large Mammals on Private Land in Natal." *Lammergeyer* 34:1–58.

Hudson, R. J. 1993. "Origins of Wildlife Management in the Western World." In *Commercialization and Wildlife Management: Dancing with the Devil,* ed. A. W. L. Hawley, 5–21. Malabar, Fla.: Krieger Publishing Co.

Hudson, R. J., K. R. Drew, and L. M. Baskin. 1989. *Wildlife Production Systems—Economic Utilization of Wild Ungulates.* Cambridge, New York, Port Chester, Melbourne, and Sydney: Cambridge University Press.

Hughes, D. J. 1984. "Meat Processors' View of Game Ranching for Meat Production." *Proceedings of the 1984 International Ranchers Roundup.* San Angelo, Texas, 336–43.

Hunter, M. L. 1989. "Aardvarks and Arcadia: Two Principles of Wildlife Research." *Wildlife Society Bulletin* 17:350–51.

Isakovic, I. 1970. "Game Management in Yugoslavia." *Journal of Wildlife Management* 34(4):800–12.

Iso-Ahola, S. E. 1980. *The Social Psychology of Leisure and Recreation.* Dubuque, Iowa: Wm. C. Brown.

Jackson, R. M., and R. K. Anderson. 1982. "Hunter-landowner Relationship: A Management and Educational Perspective." *Transactions, North American Wildlife and Natural Resources Conference* 47:693–703.

Jacobs, J. 1992. *Systems of Survival.* New York: Random House.

Jahn, L. R. 1990. "The Future of Access to Private Lands." In *R. D. No. 740,* ed. W. N. Grafton, A. Ferrise, D. Colyer, D. K. Smith, and J. E. Miller, 3–8. West Virginia University Extension Service. Morgantown, West Virginia.

Johnson, H. D. 1966. A study of organized efforts to improve landuser-sportsman relations for the purpose of maintaining public upland game hunting. Master's thesis, Utah State University, Logan.

Jordan, G., assoc. ed. 1989. "New Landowner Compensation Sought by DOW." *News for Landowners* 7(4):1–2.

Joubert, E. 1974. "The Development of Wildlife Utilization in Southwest Africa." *Journal of the South African Wildlife Management Association* 4(1):35–42.

Joubert, E., P. A. J. Brand, and G. P. Visagie. 1983. "An Appraisal of the Utilisation of Game on Private Land in Southwest Africa." *Madoqua* 3(13):197–219.

Kaiser, R. A., and B. A. Wright. 1985. "Recreational Use of Private Lands: Beyond the Liability Hurdle." *Journal of Soil and Water Conservation* 40:478–81.

Kallman, H., C. D. Agee, W. R. Goforth, and J. P. Linduska, eds. 1987. *Restoring America's Wildlife 1937–1987.* Washington, D.C.: USDI Fish and Wildlife Service.

Kelher, G. H. 1943. "The State-Sportsman-Landowner Triangle." *Journal of Wildlife Management* 7(1):7–10.

Kellert, S. R. 1981. "Wildlife and the Private Landowner." In *Wildlife Management on Private Lands*, ed. R. T. Dumke, G. V. Burger, and J. R. March, 18–41. Madison, Wis.: Department of Natural Resources.

Kimball, T. L. 1963. "For Public Recreation: Private Development of Hunting and Fishing." *Journal of Soil and Water Conservation* 18(2):49–53.

Kirby, S. B., K. M. Babcock, S. L. Sheriff, and D. J. Witter. 1981. "Private Land and Wildlife in Missouri: A Study of Farm Operator Values." In *Wildlife Management on Private Lands*, ed. R. T. Dumke, G. V. Burger, and J. R. March, 88–101. Madison, Wis.: Department of Natural Resources.

Klein, D. R. 1989. "Northern Subsistence Hunting Economics." In *Wildlife Production Systems*, ed. R. J. Hudson, K. R. Drew, and L. M. Baskin, 96–111. Cambridge, New York, Port Chester, Melbourne, and Sydney: Cambridge University Press.

Klussmann, W. G. 1966. "Deer and Commercialized Hunting Systems in Texas." In *Proceedings, The White-tailed Deer: Its Problems and Potentials*, 18–21. Texas A&M University, College Station.

Kramer, B. M. 1982. *Legal Aspects of Use and Development of Wildlife Resources on Private Lands: Colorado, Kansas, New Mexico, Oklahoma, Texas*. Great Plains Agriculture Council Publication 103. Texas Tech University School of Law, Lubbock.

Kruger, T. 1929. "Are We Drifting into European Systems of Game Management?" *Journal of Forestry* 27(3):262–63.

Lambrechts, A. V. W. 1976. Nantleding van die Natuurb-ewaringsovattinge van wildreservaateienaars in die oos-Transvaalse laeveld. Master of Science. University of Pretoria.

Langer, L. L. 1987. "Hunter Participation in Fee Access Hunting." *Transactions, North American Wildlife and Natural Resources Conference* 52:475–81.

LaPage, W. 1967. Camper characteristics differ at public and commercial campgrounds in New England. USDA Forest Service Research Note NE-59.

Larson, J. S. 1958. "Straight Answers about Posted Land." *Transactions, North American Wildlife and Natural Resources Conference* 24:480–87.

Latham, A. 1989. "The Cosmic Elk Factor: Who Owns New Mexico's Wildlife?" *New Mexico Resources* 2(3):5–9, 12.

Laycock, G. 1981. "Socking It to the Poachers." *Outdoor Life* January:54–56.

Laycock, W. A. 1987. "History of Grassland Plowing and Grass Planting on the Great Plains." In *Impacts of the Conservation Reserve Program in the Great Plains*, ed. J. E. Mitchell, 3–8. USDA Forest Service General Technical Report RM-158.

Lee, H. T. 1965. "Preserve Hunting." *Texas Parks and Wildlife* 23(12):8–9.

Leopold, A. 1921. "Wilderness and Its Place in Forest Recreational Policy." *Journal of Forestry* 19:718–21.

———. 1929. "Report of the Committee on American Wildlife Policy." *Transactions, American Game Conference* 16:196–210.

———. 1930. "The American Game Policy." *Transactions, American Game Conference* 17:284–307.

————. 1933. *Game Management*. New York: Charles Scribner's Sons.

Leopold, A. 1940. "The State of the Profession." *Journal of Wildlife Management* 4(3):343–46.

————. 1949. *A Sand County Almanac with Essays on Conservation from Round River*. New York: Oxford University Press.

Leopold, A., S. A. Cain, C. M. Cottam, I. N. Gabrielson, and T. L. Kimball. 1963. "Study of Wildlife Problems in National Parks." *Transactions, North American Wildlife and Natural Resources Conference* 28:28–45.

Lewis, D., and N. Carter, eds. 1993. *Voices from Africa: Local Perspectives on Conservation*. Baltimore, Md.: World Wildlife Fund.

Lewis, J. H. 1965. The role of large private forest ownership in outdoor recreation in Louisiana. Master's thesis, Louisiana State University, Baton Rouge.

Livingston, J. A. 1986. "Some Reflections on Integrated Wildlife and Forest Management." *Trumpeter* 3:24–29.

Lloyd, M., R. Kahn, L. Sikorowski, and S. Apker. 1996. "Public Hunting Evaluations of the Ranching for Wildlife Program." *Human Dimensions of Wildlife* 1(14):82–83.

Lucas, R. C. 1964. *User Concepts of Wilderness and their Implications for Resource Management*. Western Resource Papers, 29–39.

————. 1985. *Visitor Characteristics, Attitudes, and Use Patterns in the Bob Marshall Wilderness Complex, 1970–82*. USDA Forest Service Research Paper INT-345.

Luxmoore, R. A. 1985. "Game Farming in South Africa as a Force in Conservation." *Oryx* 19:225–31.

————. 1989. "Impact on Conservation." In *Wildlife Production Systems*, ed. R. J. Hudson, K. R. Drew, and L. M. Baskin, 413–23. Cambridge, New York, Port Chester, Melbourne, and Sydney: Cambridge University Press.

McConnell, C. A. 1981. "Common Threads in Successful Programs Benefiting Wildlife on Private Land." In *Wildlife Management on Private Lands*, ed. R. T. Dumke, G. V. Burger, and J. R. March, 279–87. Madison, Wis.: Department of Natural Resources.

McCurdy, D. R., and H. Echelberger. 1968. "The Hunting Lease in Illinois." *Journal of Forestry* 66(2):124–27.

McKinney, Larry. 1993. "Reauthorizing the Endangered Species Act—Incentives for Rural Landowners." In *Building Economic Incentives into the Endangered Species Act*, 71–78. Washington, D.C.: Defenders of Wildlife.

Mansfield, T. J., K. E. Mayer, and R. L. Callas. 1989. "California's Private Lands Wildlife Management Area Program." *Transactions, Western Section, Wildlife Society* 25:45–48.

Markham, R. W. 1986. *Establishing a Wildlife Conservancy*. Technical guide for farmers No. 18. Pietermaritzburg, Republic of South Africa: Natal Parks Board.

Morrill, W. I. 1987. "Fee Access Views of a Private Wildlife Management Consultant." *Transactions, North American Wildlife and Natural Resources Conference* 52:530–43.

Musgrave, R. S., S. Parker, and M. Wolok. 1993. "The Status of Poaching in the United States—Are We Protecting Our Wildlife?" *Natural Resource Journal* 33:977–1014.

National Wilderness Institute. 1994. *National Wilderness Institute Resource* 5(1):24–25.

1996 National Survey of Fishing, Hunting, and Wildlife-Associated Recreation. 1997. Washington, D.C.: U.S. Department of the Interior, Fish and Wildlife Service and U.S. Department of Commerce, Bureau of the Census.

Novak, M., J. A. Baker, M. E. Obbard, and B. Mallech, eds. 1987. *Wild Furbearer Management and Conservation in North America.* Toronto: Ontario Trappers Association under the authority of the Licensing Agreement with the Ontario Ministry of Natural Resources.

Nsanjama, H. 1993. "Introduction." In *Voices from Africa: Local Perspectives on Conservation,* ed. D. Lewis and H. Carter, 1–6. Baltimore, Md.: World Wildlife Fund.

Orr, D. W. 1990. "The Question of Management." *Conservation Biology* 4:8–9.

Ortega y Gasset, J. 1972. *Meditations on Hunting.* New York: Charles Scribner's Sons.

Outdoor Recreation Resources Commission. 1962. *Wilderness Recreation—A Report on Resources, Values and Problems.* Study Report No. 3. Washington, D.C.: U.S. Government Printing Office.

Peine, J. D., ed. 1984. *Proceedings, Conference on the Management of Biosphere Reserves.* Gatlinburg, Tennessee.

Peterson, C. C., and C. R. Madsen. 1981. "Property Tax Credits to Preserve Wetlands and Native Prairie." *Transactions, North American Wildlife and Natural Resource Conference* 46:125–29.

Pollock, N. H. 1968. The English game laws in the nineteenth century. Ph.D. diss., Johns Hopkins University.

Pope, A. C. III, C. E. Adams, and J. K. Thomas. 1984. "The Recreational and Aesthetic Value of Wildlife in Texas." *Journal of Leisure Research* 16(1):51–59.

Potter, D. R., J. C. Hendee, and R. N. Clarke. 1973. "Hunting Satisfaction: Game, Guns, or Nature?" *Transactions, North American Wildlife and Natural Resources Conference* 38:220–29.

Reiger, J. F. 1975. *American Sportsmen and the Origins of Conservation.* New York: Winchester Press.

Report of the Texas Game, Fish and Oyster Commissioner. W. G. Sterett, Commissioner. 1919. Austin, Texas.

Report of the Texas Game, Fish and Oyster Commissioner. W. W. Boyd, Commissioner. 1922. Austin, Texas.

Report of the Texas Game, Fish and Oyster Commissioner. W. W. Boyd, Commissioner. 1924. Austin, Texas.

Reynolds, H. G. 1971. "Game Production and Harvest in Czechoslovakia." *Journal of Forestry* 69(19):736–40.

Rossman, B. B., and J. Ulehla. 1977. "Psychological Reward Values Associated with Wilderness Use: A Functional Reinforcement Approach." *Environmental Behavior* 9:41–66.

Rounds, R. C. 1975. *Public Access to Private Land for Hunting.* Colorado Division of Wildlife Report No. 2.

Rowe-Rowe, D. T. 1984. *Game Utilisation on Private Land in Natal.* Pietermaritzburg, Republic of South Africa: Natal Parks Board.

———. 1985. *Activities of the Wildlife Extension Services Sub-section of the Natal Parks Board.* Pietermaritzburg, Republic of South Africa: Natal Parks Board.

Salwasser, H. 1990. "Sustainability as a Conservation Paradigm." *Conservation Biology* 4:213–16.

Sampson, N. R. 1990. "Challenges and Opportunities for Natural Resource Programs to Assist Private Landowners." In *R. D. No. 740,* ed. W. N. Grafton, A. Ferrise, D. Colyer, D. K. Smith, and J. E. Miller, 9–16. West Virginia University Extension Service. Morgantown, West Virginia.

Sargent, F. O., C. C. Boykin, O. C. Wallmo, and E. H. Cooper. 1958. "Land for Hunters . . . a Survey of Hunting Leases." *Texas Game and Fish* 16(9):22–29.

Schreye, R., and J. W. Roggenbuck. 1978. "The Influence of Experience Expectations on Crowding Perceptions and Social-Psychological Carrying Capacities." *Leisure Science* 1(4):373–93.

Scott, N. R. 1974. "Toward a Psychology of Wilderness Experience." *Natural Resources Journal* 14:231–37.

Shelton, L. R. 1969. Economic aspects of wildlife management programs on large private land holdings in the Southeast. Master's thesis, Mississippi State University, Starkville.

———. 1978. Values of big game to western slope ranchers of Colorado. Ph.D. diss., Colorado State University, Fort Collins.

———. 1982. "Constraints on Development for Wildlife on Private Lands." *Transactions, North American Wildlife and Natural Resources Conference* 47:464–69.

———. 1987. "Fee Hunting Systems and Important Factors in Wildlife Commercialization on Private Lands." In *Valuing Wildlife—Economic and Social Preferences*, ed. D. J. Decker and G. R. Goff, 109–16. Boulder, Colo.: Westview Press.

Shilling, C. L., and R. L. Bury. 1973. "Attitudes Toward Recreational Development Potentials of Non-corporate Landowners." *Journal of Leisure Research* 5(2):74–82.

Sigler, W. G. n.d. *Wildlife Law Enforcement.* 3rd edition. Dubuque, Iowa: Wm. C. Brown Company Publishers.

Smith, J. L. D., A. H. Berner, F. J. Cuthbert, and J. A. Kitts. 1992. "Interest in Fee Hunting by Minnesota Small Game Hunters." *Wildlife Society Bulletin* 20:20–26.

Smith, R. J. 1981. "Resolving the Tragedy of the Commons by Creating Private Property Rights in Wildlife." *Cata. J* 1(2):439–68.

Stankey, G. H. 1973. *Visitor Perception of Wilderness Recreation Carrying Capacity.* USDA Forest Service Research Paper INT-142.

Steele, N. A. 1981. "The Conservancy Concepts in Natal." *Nyala News* 4(50):32–34.

Steinbach, D. W., J. R. Conner, M. K. Glover, and J. M. Inglis. 1987. "Economic and Operational Characteristics of Recreational Leasing in the Edwards Plateau and Rio Grande Plains in Texas." *Transactions, North American Wildlife and Natural Resources Conference* 52:496–515.

Swank, W. G. 1981. "Wildlife Laws and their Effect on Deer Management." *Proceedings of the 1981 International Ranchers Roundup.* Uvalde, Texas, 361–66.

Taylor, R. D. 1994. "Wildlife Management and Utilization in Zimbabwean Communal Land: A Preliminary Evaluation in Nyaminyami District, Kariba." In *Wildlife Ranching: A Celebration of Diversity,* ed. W. van Hoven, H. Ebedes, and A. Conroy, 282–94. Pretoria, Republic of South Africa: Promedia.

[Teer, J. G.] 1987. "Wildlife and Habitat." In *World Resources 1987. A Report by: The International Institute for Environment and Development and the World Resources Institute,* 77–92.

Teer, J. G. 1989. "Conservation Biology—A Book Review." *Wildlife Society Bulletin* 17:337–39.

Teer, J. G., and N. K. Forrest. 1968. "Bionomic and Ethical Implications of Commercial Game Harvest Programs." *Transactions, North American Wildlife and Natural Resources Conference* 33:192–204.

Teer, J. G., G. V. Burger, and C. Y. Deknatel. 1983. "State-supported Habitat Management and Commercial Hunting on Private Lands in the United States." *Transactions, North American Wildlife and Natural Resources Conference* 48:445–56.

Thomas, J. K., C. E. Adams, and C. E. Pope III. 1984. *White-tail Deer Hunting in Texas: Socioeconomic Factors AVecting Access to Land.* Departmental Technical Report No. 84-1. Texas Agricultural Experiment Station, College Station.

Thomas, J. W., and H. Salwasser. 1989. "Bringing Conservation Biology into a Position of Influence in Natural Resources Management." *Conservation Biology* 3:123–27.

Thowardson, N. 1979. Landowner constraints on Oklahoma hunting opportunities. Master's thesis, Oklahoma State University, Stillwater.

Tober, J. A. 1981. *Who Owns the Wildlife? The Political Economy of Conservation in Nineteenth Century America.* Westport, Conn.: Greenwood Press.

Tolba, M. K. 1990. "Building an Environmental Institutional Framework for the Future." *Environmental Conservation* 17:105–10.

Train, R. E. 1978. "Who Owns American Wildlife?" In *Wildlife and America*, ed. H. P. Brokaw, 275–78. Washington, D.C.: U.S. Council on Environmental Quality.

Trefethen, J. B. 1975. *An American Crusade for Wildlife.* New York: Winchester Press.

Tucker, W. J. 1933. "Is It Wise To Tinker with the Legal Status of Game?" In *Proceedings, 25th Conference of the International Association of Game Fish Conservation Commissions* 25:125–30.

———. 1943. "Shooting Preserves Pay Off." *Texas Game Fish* 1(4):6–7, 17.

U.S. General Accounting Office. 1989. *Wilderness Preservation: Problems in Some National Forests Should Be Addressed.* GAO/RCED-89-202. Washington, D.C.

Vaske, J. J., M. P. Donnally, T. A. Heberlein, and B. Shelby. 1982. "Differences in Reported Satisfaction Ratings by Consumptive and Nonconsumptive Recreationists." *Journal of Leisure Research* 14(3):195–206.

Vento, B. F. 1989. "A Wilderness Revolution for the 1990s Commemorating the Wilderness Act on its 25th Anniversary." In *Managing America's Enduring Wilderness Resource*, ed. D. W. Lime, 9–17. Saint Paul: University of Minnesota.

Vogeler, I. 1977. "Farm and Ranch Vacationing." *Journal of Leisure Research* 9(4):291–300.

Von Kerckerinck, J. 1987. *Deer Farming in North America.* Rhinebeck, N.Y.: Phanter Press.

Vrana, V. K. 1990. "Conservation in the '90s." *Journal of Soil and Water Conservation* 45:506.

Wagner, F. H. 1989. "American Wildlife Management at the Crossroads." *Wildlife Society Bulletin* 17:354–60.

Webb, W. L. 1960. "Forest Wildlife Management in Germany." *Journal of Wildlife Management* 24(2):147–61.

Western, D. 1991. "Biology and Conservation: Making the Relevant Connection." *Conservation Biology* 5:431–33.

White, R. J. 1987. *Big Game Ranching in the United States.* Mesilla, N.Mex.: Wild Sheep and Goat International.

Wiggers, E. P., and W. A. Rootes. 1987. "Lease Hunting: Views of the Nation's Wildlife Agencies." *Transactions, 52nd North American Wildlife and Natural Resources Conference* 52:525–29.

Wigley, T. B., and M. A. Melchiors. 1987. "State Wildlife Management Programs for Private Lands." *Wildlife Society Bulletin* 15:580–84.

Wolfe, M. L. 1991. "An Historical Perspective on the European System of Wildlife Management." In *Wildlife Production: Conservation and Sustainable Development*, ed. L. Renecker and R. J. Hudson, 163–68. Agricultural and Forestry Experiment Station misc. pub. 91-6. Fairbanks: University of Alaska.

Wright, B. A., and R. A. Kaiser. 1986. "Wildlife Administrators' Perceptions of Hunter Access Problems: A National Overview." *Wildlife Society Bulletin* 14(1):30–35.

Wright, B. A., R. A. Kaiser, and J. E. Fletcher. 1988. "Hunter Access Decisions by Rural Landowners: An East Texas Example." *Wildlife Society Bulletin* 16:152–58.

Wright, R. M., and D. Western. 1994. *Natural Connections: Perspectives on Community Based Conservation.* Washington, D.C.: Island Press.

Yorks, T. P. 1989. "Ranching Native and Exotic Ungulates in the United States." In *Wildlife Production Systems: Economic Utilisation of Wild Ungulates,* ed. R. J. Hudson, K. R. Drew, and L. M. Baskin, 268–85. Cambridge: Cambridge University Press.

Young, E. 1985. "Game Farming—A Young Giant." *Natura* 5:3.

Young, R. A., and R. Crandall. 1984. "Wilderness Use and Self-actualization." *Journal of Leisure Research* 16(2):149–60.

Index

Pages containing illustrations appear in italics.